SpringerBriefs in Materials

For further volumes:
http://www.springer.com/series/10111

Chirag R. Gajjar · Martin W. King

Resorbable Fiber-Forming Polymers for Biotextile Applications

 Springer

Chirag R. Gajjar
Martin W. King
College of Textiles
North Carolina State University
Raleigh, NC
USA

ISSN 2192-1091   ISSN 2192-1105   (electronic)
ISBN 978-3-319-08304-9   ISBN 978-3-319-08305-6   (eBook)
DOI 10.1007/978-3-319-08305-6
Springer Cham Heidelberg New York Dordrecht London

Library of Congress Control Number: 2014942305

Printed on acid-free paper

Springer is part of Springer Science+Business Media (www.springer.com)

Preface

Over the past 50 years, the quest for materials intended to interface with biological systems in order to evaluate, treat, augment, or replace any tissue, organ or function of the body has resulted in the creation of the interdisciplinary field called *Biomaterials Science*. Then in the 1960s, with the introduction of the first synthetic resorbable suture Dexon®, the new field of resorbable biomaterials was launched. Since that time, a new generation of synthetic bioresorbable polymers has been developed specifically for biomedical applications, and there are a number of driving forces for the favorable consideration of bioresorbable over biostable materials. They include: (i) improved long-term biocompatibility compared to many of the existing permanent implantable materials, (ii) the advantage of not having to remove the bioresorbable implant once it is no longer needed, thereby avoiding a second surgical operation, and (iii) the growth of emerging biomedical technologies, such as tissue engineering, regenerative medicine, gene therapy, controlled drug delivery, and bionanotechnology, all of which require bioresorbable materials. It can be predicted that in the near future, bioresorbable devices that will assist in the repair and regeneration of damaged, diseased, and injured tissues will replace many of the permanent prosthetic devices. Thus, the topic of *Bioresorbable Biomaterials* generates considerable interest and research activity in the field of biomaterials science today.

Because of the structural requirements of medical devices for their intended end-use and the technological advancements in synthetic fibers and textile technology, a new field of *Biotextiles* has evolved to utilize the potential of various woven, knitted, braided, and nonwoven textile structures for biomedical applications. Biotextiles are defined as those structures composed of textile fibers designed for use in specific biological environments where their performance depends on their biocompatibility and biostability with cells and biological fluids. Textile structures consisting of bioresorbable fibers provide certain unique properties to the medical device. One example for instance, involves the filaments of a braided suture, which initially provides the strength and flexibility to hold the wound together. A second example addresses the dimensional stability of a textile tissue engineering scaffold, which maintains its required shape and size while withstanding the initial shear forces imposed by the circulating culture media in a bioreactor, the contractile forces imposed by growing cells, and the compressive forces from the surrounding tissue. In addition, because textile structures have an inherently high level of porosity, they can encourage cell growth and promote migration and proliferation. Thus, most of the applications of bioresorbable polymer devices such as sutures and scaffolds require the biomaterial to be fabricated into a dimensionally stable structure so as to function effectively. For these reasons, the use of bioresorbable polymers as fibers is currently dominating the field of resorbable biomaterials for biomedical applications.

This review focuses on those synthetic bioresorbable polymers that can be spun into fibers or filaments, and subsequently used as biotextiles. We have listed and reported on the properties and applications of both conventional and commercially available fiber-forming bioresorbable polymers as well as those that are still being developed experimentally. Factors affecting the performance of these biomaterials are presented and the precautionary measures that may be taken to reduce the hydrolytic degradation during manufacturing and processing are discussed.

Chirag R. Gajjar
Martin W. King

Contents

Chapter 1
Overview of Resorbable Biomaterials

Keywords Bioabsorption · Biocompatibility · Biodegradation · Bioerosion · Biofunctionality · Biomaterial · Bioresorption

1.1 Introduction

A 'biomaterial' is defined as a material intended to interface with biological systems to evaluate, treat, augment or replace any tissue, organ or function of the body [1]. The history of using various natural biomaterials for different applications dates back several thousands of years. As early as 3500 B.C., linen, silk, flax, hair, grass, and animal gut have been found to be used as suture materials in ancient Egyptian and Indian civilizations [2, 3]. Over time, many different natural and synthetic polymers have been explored as biomaterials for various medical applications. However, in the last few decades, there has been a paradigm shift from biostable biomaterials to biodegradable biomaterials [4, 5]. With the introduction of the first synthetic resorbable suture Dexon® in 1960s, a new field of bioresorbable biomaterials has evolved, and since then, there has been rapid development of a new generation of synthetic biodegradable polymers specifically developed for biomedical applications.

There are some major driving forces for the favorable consideration of bioresorbable over biostable materials for biomedical applications. Firstly, there are long-term biocompatibility issues with many of the existing permanent implants. Secondly, there is an advantage of not having to remove the bioresorbable implant once it is no longer needed, thereby avoiding the issue related to repeat surgery. Finally, there are emerging novel biomedical technologies such as tissue engineering, regenerative medicine, gene therapy, controlled drug delivery, and bionanotechnology, all of which demand bioresorbable materials [6]. With the current trend, it can be predicted that in the near future, bioresorbable devices that

C. R. Gajjar and M. W. King, *Resorbable Fiber-Forming Polymers for Biotextile Applications*, SpringerBriefs in Materials, DOI: 10.1007/978-3-319-08305-6_1, © The Author(s) 2014

could help in the repair and regeneration of damaged tissues will replace many of the permanent prosthetic devices. Thus, it seems logical that *'Bioresorbable Biomaterials'* is the topic of contemporary interest in the field of biomaterials today.

Before we delve further in this area, there are certain considerations that should be kept in mind. The field of bioresorbable polymers is relatively new. Complete understanding of the structure-property relationship for bioresorbable polymers is yet to be achieved. Chemists, material scientists, engineers, and biologists need to collaborate to unveil the mysteries of resorbable biomaterial design. Again, this field is very wide and a single, ideal polymer does not exist. Instead, polymers from a variety of sources have to be synthesized and engineered to best match the desired biomedical function and required resorption rate. Another issue is that the word 'bioresorption' is loosely defined and many times used interchangeably with words like biodegradation, bioabsorption, and bioerosion. Similarly, the use of the prefix 'bio' is not well established, leading to interchangeable use of terms 'resorbable' and 'bioresorbable' or 'degradable' and 'biodegradable'. Despite efforts made to establish widely accepted definitions of these terms [7], there is still significant confusion even among the experts regarding the usage of these terminologies. The term 'biodegradation' refers to the chemical degradation of the material caused by biological agents like enzymes or microorganisms (for example, the degradation of plastics in landfills). The term 'bioerosion' refers to the physical change in size, shape, or mass of a material caused either by degradation or simply dissolution in biological conditions. 'Bioresorption' implies that the polymer or its degradation products (formed by simple hydrolysis) are removed from the body by cellular activity (such as phagocytosis or the citric acid cycle) in a biological environment and is the correct technical term to use in the case of resorbable medical devices [8].

1.2 Requisites for Resorbable Biomaterials

Despite its debut in the late 1960s and large amounts of research initiated thereafter, the field of synthetic bioresorbable polymers has been slow to evolve. There are some unique challenges associated with bioresorbable biomaterials. In order to qualify as a biomaterial, a material must demonstrate biocompatibility, i.e., the ability to perform with an appropriate host response in a specific application. In addition to this, a bioresorbable biomaterial must demonstrate biocompatibility over a period of time. There are changes in the chemical, physical, mechanical, and biological properties of bioresorbable biomaterials induced by degradation over time. These changes can cause different host responses than the original parent material [9]. Hence, the following list includes some important properties which must be considered in the design of bioresorbable biomaterials [10].

1.2.1 Biocompatibility

- Nontoxic
- Non-inflammatory
- Non-immunogenic
- Non-carcinogenic
- Non-thrombogenic

1.2.2 Biofunctionality

- Adequate mechanical properties over time
- Appropriate permeability and processability for intended application
- Appropriate rate of degradation to match the healing/regeneration process
- Elimination of residues/degradation products safely
- Resistance to sterilization
- Long shelf life
- Approval by regulatory agencies for use in specific applications

Because of such complex requirements, the number of synthetic bioresorbable polymers for practical use is limited. This must have been the limitation faced by nature too since all life on earth is based on only three polymeric backbones, namely, polynucleotide, poly(α-amino acid), and polysaccharide chains [11]. For the same reasons, synthetic bioresorbable polymers have been restricted only to certain classes of polymers and till now only a few of them have been approved by the FDA for certain specific applications.

Given the complexities and restrictions with synthetic bioresorbable polymers, one might tend to prefer natural polymers which have structural similarity to components in host tissues. Natural polymers also possess several inherent advantages such as bioactivity, the ability to present receptor-binding ligands to cells, susceptibility to cell-triggered proteolytic degradation, and natural remodeling [6]. However, strong inherent bioactivity of these natural polymers has its own downsides such as strong immunogenic response and the potential for rejection. Other disadvantages include insufficient mechanical strength, difficulty in obtaining consistent quality, the need for purification, and the possibility of disease transmission [12]. Synthetic biomaterials on the other hand are usually biologically inert and have more predictable properties and batch-to-batch uniformity. Their properties could be tailored for specific applications. For these reasons, synthetic polymers have been the attractive choice for bioresorbable devices.

The earliest synthetic bioresorbable polymers were based on linear aliphatic polyesters. More recently, many other classes of polymers have been explored to develop new resorbable systems for various biomedical applications. Today, a variety of polymer systems having tailored bioresorption profiles are being used or researched for specific biomedical applications. All these bioresorbable polymeric biomaterials have been divided into the following eight groups based on their chemical origin [13]:

1. Bioresorbable linear aliphatic polyesters (e.g., polyglycolide, polylactide, polycaprolactone, polyhydroxybutyrate) and their copolymers within the aliphatic polyester family like poly(glycolide-L-lactide) copolymer and poly(glycolide-ε-caprolactone) copolymer.
2. Bioresorbable copolymers between linear aliphatic polyesters in (1) and monomers other than linear aliphatic polyesters like, poly(glycolide-trimethylene carbonate) copolymer, poly(L-lactic acid-L-lysine) copolymer, tyrosine-based polyarylates or polyiminocarbonates or polycarbonates, poly(D,L-lactide-urethane), and poly(ester-amide).
3. Polyanhydrides.
4. Poly(orthoesters).
5. Poly(ester-ethers) like poly-*p*-dioxanone.
6. Bioresorbable polysaccharides like hyaluronic acid, chitin, and chitosan.
7. Polyamino acids like poly-L-glutamic acid and poly-L-lysine.
8. Inorganic bioresorbable polymers like polyphosphazene and poly [bis (carboxylatophenoxy) phosphazene] which have a nitrogen–phosphorus backbone instead of an ester linkage.

References

1. D.F. Williams, *The Williams Dictionary of Biomaterials* (Liverpool University Press, Liverpool, 1999)
2. S.W. Shalaby, K.J.L. Burg, *Absorbable and Biodegradable Polymers* (CRC Press, Boca Raton, 2004)
3. H. Planck, M. Dauner, M. Renardy (eds.), *Medical Textiles for Implantation* (Springer, Berlin, 1990)
4. A.J. Domb, J. Kost, D.M. Wiseman, *Handbook of Biodegradable Polymers* (Harwood Academic Publishers, Newark, 1997)
5. E. Pişkin, Biodegradable polymers as biomaterials. J. Biomater. Sci. Polym. Ed. **6**(9), 775–795 (1995)
6. L.S. Nair, C.T. Laurencin, Biodegradable polymers as biomaterials. Prog. Polym. Sci. **32**(8–9), 762–798 (2007)
7. D.F. Williams, European Society for Biomaterials, in *Definitions in Biomaterials: Proceedings of a Consensus Conference of the European Society for Biomaterials*, Chester, England, 3–5 March 1986 (Elsevier, 1987)
8. J. Kohn, S. Abramson, R. Langer, Bioresorbable and bioerodible materials, in *Biomaterials Science—An Introduction to Materials in Medicine*, 2nd edn., ed. by B.D. Ratner, A.S. Hoffman, F.J. Schoen, J.E. Lemons (Elsevier, Amsterdam, 2004), pp. 115–127
9. B.D. Ulery, L.S. Nair, C.T. Laurencin, Biomedical applications of biodegradable polymers. J. Polym. Sci., Part B: Polym. Phys. **49**(12), 832–864 (2011)
10. M. Vert, Polymeric biomaterials: strategies of the past vs. strategies of the future. Prog. Polym. Sci. **32**(8–9), 755–761 (2007)
11. M. Vert, Aliphatic polyesters: great degradable polymers that cannot do everything. Biomacromolecules **6**(2), 538–546 (2005)

12. T. Hayashi, Biodegradable polymers for biomedical uses. Prog. Polym. Sci. **19**(4), 663–702 (1994)
13. C.C. Chu, Biodegradable polymeric biomaterials: an updated overview, in *Biomedical Engineering Handbook*, 2nd edn., ed. by J.D. Bronzino (CRC Press, Boca Raton, 2000), pp. 1–22

Chapter 2
Degradation Process

Keywords Degradation mechanism · Hydrolytic degradation · Autocatalytic degradation · Bulk erosion · Surface erosion · Strength and mass loss profile · Critical device dimension · Hydrolysis

2.1 Degradation Mechanism for Bioresorbable Polymers

A typical resorption profile of a material consists of four steps, namely water sorption, reduction of mechanical properties, reduction of molar mass, and finally complete loss of weight, as shown in (Fig. 2.1) [1]. Initially, water and/or biological fluids diffuse into the material, followed by a reduction in the mechanical properties. Loss in strength and modulus is initially due to the plasticizing effect of the fluids and later due to the reduction in molar mass. Change in shape and weight loss are the final stages before the material is completely resorbed [2].

Different resorbable polymers have different strength and mass loss profiles. However, all of them have one common characteristic, i.e., strength loss always occurs much earlier than mass loss. Initially, hydrolysis takes place in the amorphous regions of the polymer, converting long polymer chains into shorter fragments. This causes a reduction in molecular weight without a loss in physical properties as the matrix is still held together by the crystalline regions. Reduction in molecular weight is soon followed by a reduction in physical and mechanical properties as more and more chains are converted to smaller fragments. At the later stages, the fragments are metabolized and converted to oligomers which are either absorbed or excreted by the body. The metabolization of the fragments results in a rapid mass loss for the polymer [3, 4].

The process of bioresorption can be initiated by any of the following four ways:

C. R. Gajjar and M. W. King, *Resorbable Fiber-Forming Polymers for Biotextile Applications*, SpringerBriefs in Materials, DOI: 10.1007/978-3-319-08305-6_2, © The Author(s) 2014

Fig. 2.1 Typical changes in the properties during bioresorption of a material [2]

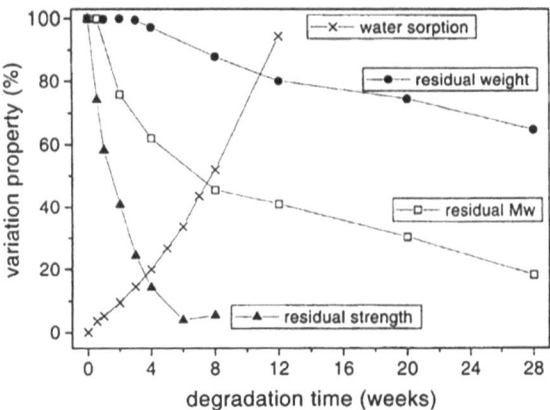

Fig. 2.2 Hydrolysis of a chemical bond

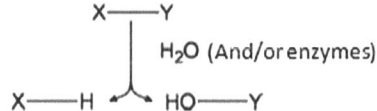

1. **Solubilization** (e.g., dextran, polyvinyl alcohol, and polyethylene oxide);
2. **Ionization followed by solubilization** (e.g., polyacrylic acid and polyvinyl acetate);
3. **Enzymatically catalyzed hydrolysis** (e.g., polysaccharides and polyamides);
4. **Simple hydrolysis** (e.g., aliphatic polyesters).

Simple hydrolysis and enzymatically induced hydrolysis are the two main mechanisms for bioresorption of various resorbable implantable materials. Both the mechanisms involve cleavage of bonds that are susceptible to hydrolysis in the polymer backbone, causing fragmentation of the whole polymeric structure and producing low molecular weight oligomers that can be absorbed or excreted by the body, as shown in Fig. (2.2).

2.2 Bulk Erosion Versus Surface Erosion

The resorption process of a biomaterial is also classified according to its erosion mechanism [5]. Bulk erosion is the mechanism when water diffuses rapidly into a polymer structure, leading to hydrolysis. The subsequent mass loss then occurs throughout the bulk of the material, as shown in (Fig. 2.3c). In this case, the rate of diffusion of water into the substrate is higher than the rate of hydrolysis. And as a result, the water that penetrates the substrate leads to hydrolysis from the inside out. A characteristic behavior of bulk eroding polymers is the sudden and rapid loss of strength and structural integrity as the resorption continues over time.

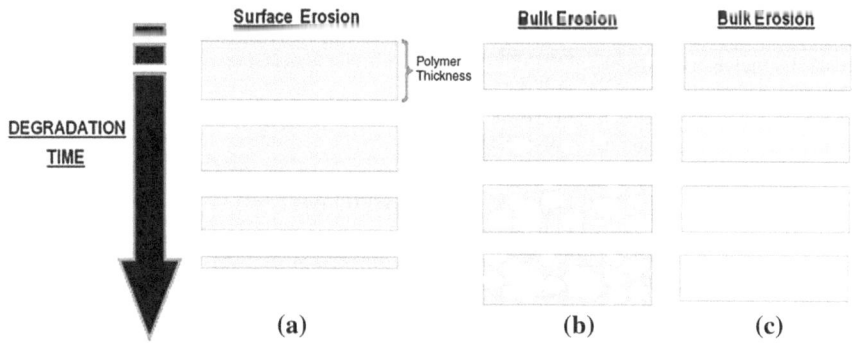

Fig. 2.3 Schematic illustration of erosion mechanisms: **a** surface erosion, **b** bulk erosion with autocatalysis, and **c** bulk erosion without autocatalysis [10]

Another mechanism is surface erosion, in which the mass loss occurs at the water/implant interface, causing the implant to resorb from its outer surface toward its center while maintaining its bulk integrity. This is sometimes referred to as 'device thinning,' as shown in (Fig. 2.3a). In this case, the rate of hydrolysis is higher than the rate of diffusion of water. Most of the enzymatically resorbable polymers show surface erosion. Surface-eroding polymers have a better ability to achieve zero-order release kinetics and are therefore ideal candidates for developing drug delivery devices [6].

It has been proposed that all resorbable polymers can undergo both surface erosion and bulk erosion. The option followed in a particular situation is governed not only by the rate of hydrolysis of the functional groups in the backbone chain and the rate of diffusion of water inside the matrix, but also by the dimensions of the matrix. A critical device dimension ($L_{critical}$) can be calculated for each resorbable polymer. If the thickness of a matrix is larger than $L_{critical}$, then it will undergo surface erosion. However, if it is smaller, then bulk erosion will occur [7]. $L_{critical}$ values for some of the resorbable polymers are given in Table 2.1.

Another factor that complicates the resorption behavior of resorbable polymers is the phenomenon called autocatalytic degradation [8]. Autocatalysis occurs when the reaction product is itself the catalyst for the same reaction. In the case of bulk eroding polymers, the oligomeric hydrolysis products (usually carboxylic and other acids) are retained within the material, causing a localized decrease in pH which accelerates the rate of degradation [5]. As a result of this self-catalyzed hydrolysis, hollow structures are formed within the polymer, which leads to a rapid deterioration of the mechanical properties and sudden loss of structural integrity, as shown in (Fig. 2.3b). For example, in the case of poly(lactide) (PLA), when the thickness of the substrate is larger than 2 mm, the entrapment and accumulation of acidic oligomers and monomer result in autocatalytic degradation [9]. It should be noted that even though polymers such as PLA and poly(lactide-co-glycolide) (PLGA) are being widely studied as scaffold materials for tissue regeneration, the by-products and the acidic condition generated by the rapid autocatalytic

Table 2.1 Critical thickness ($L_{critical}$) values for selected resorbable polymers above which the hydrolytic resorption mechanism changes from bulk erosion to surface erosion [7]

Polymer	$L_{critical}$
Poly(anhydride)	75 μm
Poly(ketal)	0.4 mm
Poly(ortho esters)	0.6 mm
Poly(acetal)	2.4 cm
Poly(ε-caprolactone)	1.3 cm
Poly(α-hydroxy esters) (PLA)	7.4 cm
Poly(amides)	13.4 m

degradation of these polymers might be detrimental for the growth of certain cell lines. The effect of an autocatalytic degradation mechanism on the growth of cells should be considered when selecting the resorbable scaffold material.

References

1. R.L. Kronenthal, Biodegradable polymers in medicine and surgery, in *Polymers in Medicine and Surgery*, ed. by R.L. Kronenthal, Z. Oser, E. Martin (Plenum Press, Berlin, 1975), pp. 119–137
2. L. Fambri, C. Migliaresi, K. Kesenci, E. Piskin, Biodegradable Polymers, in *Integrated Biomaterials Science*, ed. by R. Barbucci (Kluwer Academic, Plenum, Berlin, 2002), pp. 119–187
3. W.S. Pietrzak, D.R. Sarver, M.L. Verstynen, Bioabsorbable polymer science for the practicing surgeon. J. Craniofac. Surg. **8**(2), 87–91 (1997)
4. J.C. Middleton, A.J. Tipton, Synthetic biodegradable polymers as orthopedic devices. Biomaterials **21**(23), 2335–2346 (2000)
5. A. Göpferich, Mechanisms of polymer degradation and erosion. Biomaterials **17**(2), 103–114 (1996)
6. A. Göpferich, J. Tessmar, Polyanhydride degradation and erosion. Adv. Drug Deliv. Rev. **54**(7), 911–931 (2002)
7. F. von Burkersroda, L. Schedl, A. Göpferich, Why degradable polymers undergo surface erosion or bulk erosion. Biomaterials **23**(21), 4221–4231 (2002)
8. G.L. Siparsky, K.J. Voorhees, F. Miao, Hydrolysis of polylactic acid (PLA) and polycaprolactone (PCL) in aqueous acetonitrile solutions: autocatalysis. J. Polym. Environ. **6**(1), 31–41 (1998)
9. S. Li, M. Vert, Biodegradation of aliphatic polyesters, in *Degradable Polymers: Principles and Applications*, 2nd edn., ed. by G. Scott (Springer, Netherlands, 2002), pp. 71–131
10. A.C. Vieira, J.C. Vieira, J.M. Ferra, F.D. Magalhães, R.M. Guedes, A.T. Marques, Mechanical study of PLA–PCL fibers during in vitro degradation. J. Mech. Behav. Biomed. Mater. **4**(3), 451–460 (2011)

Chapter 3
Biotextiles: Fiber to Fabric for Medical Applications

Keywords Biotextile · Fibers and filaments · Melt spinning · Bicomponent spinning · Textile structures · Braids · Knits · Nonwovens · Spacer fabric

3.1 Importance of Fiber-Forming Biopolymers

Among the medical devices made out of bioresorbable polymers, fibrous materials are playing an increasingly important role. This is attributed mainly to the structural requirements of medical devices for the intended end use and the technological advancements made in synthetic fibers and textile technology. A new field of 'biotextiles' has evolved to utilize the potential of various nonwoven, woven, knitted and braided textile structures for biomedical applications. The term *'biotextile'* is defined as the structure composed of textile fibers designed for use in a specific biological environment as a medical device for the prevention, diagnosis, or treatment of an injury or disease, and whose performance in improving the health and wellness of the patient depends on both its biocompatibility and biostability with cells and biological fluids [1].

The field of biotextiles lies at the intersection of the disciplines of polymer and fiber science, textile technology, biomedical engineering, surface science, biomechanics, cell biology as well as human anatomy and physiology. Hence, each of these fields has an important role to play. Textile structures consisting of bioresorbable fibers provide certain unique properties to medical devices. For example, the filaments in a braided dimensionally stable suture provide the initial strength and flexibility to hold the margins of a wound together while it heals. In another example, a textile tissue engineering scaffold can maintain its desired shape and size while withstanding the shear forces imposed by the circulating culture media in a bioreactor, the contractile forces imposed by the growing cells, as well as the other compressive forces from the surrounding tissues. In addition, textile structures have an inherent and significant level of porosity, which

C. R. Gajjar and M. W. King, *Resorbable Fiber-Forming Polymers for Biotextile Applications*, SpringerBriefs in Materials, DOI: 10.1007/978-3-319-08305-6_3, © The Author(s) 2014

Table 3.1 Unique advantages of fiber-based biotextiles

Fiber-based biotextiles can provide
• Thin, strong, flexible, lightweight, and porous structures
• Substrates with excellent fatigue properties
• A very large surface area desirable for drug delivery and cell attachment
• A structure that has the ability to be folded or compressed into a small volume for less invasive delivery through a catheter
• A structure that has the ability to tolerate repeated needle penetration and other types of iatrogenic damage (iatrogenic refers to any adverse condition induced in a patient by a physician's activity, manner, or therapy)
• A structure that encourages the infiltration and proliferation of cells for improved tissue regeneration and biocompatibility

promotes cell growth and proliferation. Thus, most of the applications for bioresorbable polymer devices, such as sutures and scaffolds, require bioresorbable polymers to be fabricated into dimensionally stable textile structures for their effective performance [2–4]. Table 3.1 lists the unique advantages presented by fiber-based biotextiles. For these reasons, the use of fiber-forming bioresorbable polymers is currently dominating the field of bioresorbable biomaterials for biomedical applications.

3.2 Requisites for Fiber-Forming Bioresorbable Polymers

Theoretically, all the polymers that can form a viscous melt or a solution can be converted into filaments or fibers. For example, even a low molecular weight sugar solution can be spun into fibers to make 'cotton candy,' but the resulting fibers do not have adequate stability or mechanical properties for biotextile applications. Thus, there are certain prerequisites which limit the number of polymers that can be practically extruded into fibers or filaments for various end uses. The thermal, mechanical, and physical properties of a fiber depend largely on the chemical structure of the polymer, which governs the rate and extent of crystallization including the shape, dimensions, and orientation of the crystals which are formed during cooling. These factors have a great impact on the ability of a polymer to form fibers. A fiber-forming polymer should have a chemical structure that allows it to be spun into filaments that are predisposed to form an oriented crystalline structure on stretching or drawing [5].

A fiber-forming polymer should have a high molecular weight and long molecular chain length, i.e., the degree of polymerization should be sufficiently high. The minimum molecular weight required for fiber-forming properties varies with the chemical nature of the polymer. In general, the lower the interchain cohesive forces, the higher is the minimum molecular weight required for fiber

Table 3.2 Requisites for fiber-forming polymers

Preferable conditions for fiber formation
• Intermediate to high molecular weight (approximately 20,000–250,000 Da)
• Linear polymer chains without bulky side groups, cross-links, or side chains to facilitate rapid crystallization on cooling from the melt
• The polymer should be readily soluble in a solvent, or it should be capable of melt processing
• The level of intermolecular bonding or the chain entanglement should be such that the melt viscosity is not too high or too low
• Melting temperature should not be close to the decomposition temperature

formation. The crystallinity of a polymer affects its tensile modulus, and to some extent, its tensile strength. Therefore, in order to have fibers with high tensile modulus and strength, the molecular chains must be easily crystallizable, and there should be no localized segmental mobility or overall chain mobility. Usually, linear polymers without bulky side groups have this capability. Bulky side groups in the polymer chain reduce the ease of crystallization. Moreover, if there are reactive side chains, they tend to result in a cross-linked, three-dimensional polymer network. Such polymers are insoluble, infusible gels, or rubbers and cannot be spun into functional fibers [6].

In order to extrude fibers or filaments, a polymer should either be readily soluble in a solvent or it should be capable of melt processing. In the latter case, the melting temperature (Tm) of the polymer should not be so high as to affect the processability and it should be lower than its decomposition temperature. However, in order to have the thermal stability, the Tm should be much higher than the temperature it will be exposed to during normal use. Depending upon the end use, the fibers should have the required modulus, rigidity, or stiffness. The glass transition temperature (Tg) of a polymer plays an important role in making the fibers stiff or flexible at the normal 'use' temperature. Thus, the Tg and Tm play important roles in determining the processability of fiber-forming polymers and their potential end use applications.

Given that such prerequisites need be met in order to ensure fiber formation, it is evident that not all bioresorbable polymers are fiber forming. One therefore needs to identify those bioresorbable polymers that meet the criteria summarized in Table 3.2, in order to ensure fiber formation for biomedical applications.

3.3 Manufacturing Processes for Biotextiles

In order to understand the resorption behavior of bioresorbable fibers and biotextiles when exposed to various processing parameters, it is necessary to understand the manufacturing processes for biotextiles. The majority of resorbable polymers are synthetic materials that have been engineered chemically to include hydrolysable groups. Hence, to begin with, most of these polymers are in the form

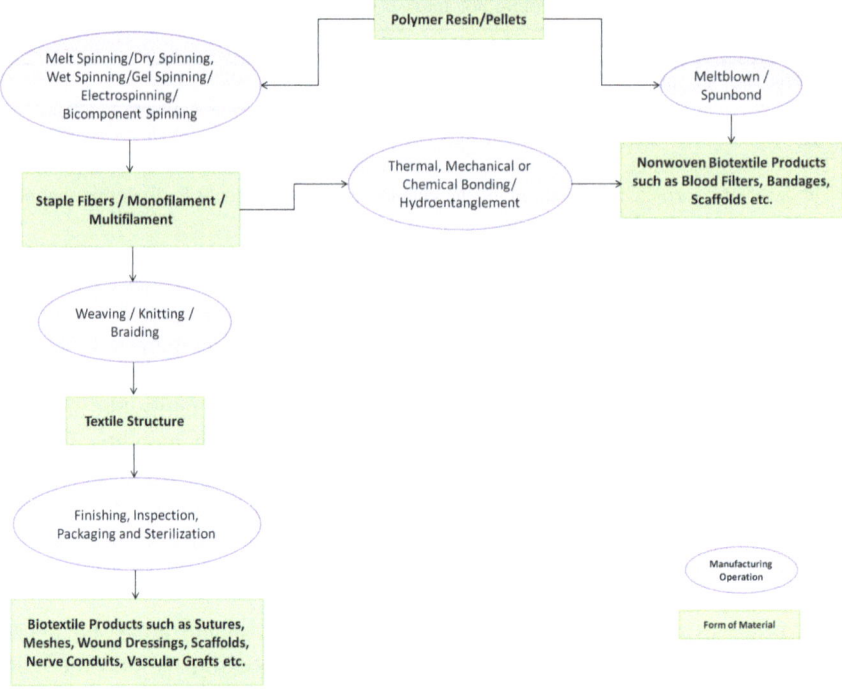

Fig. 3.1 Flow chart of typical manufacturing operations for biotextile products

of resins or pellets. These are then converted into fibers or filaments and into various textile structures using the manufacturing operations shown in Fig. 3.1.

3.3.1 Fiber Extrusion Spinning

All textile-based medical devices consist of structures made from textile fibers. What is a textile fiber, and how is it distinguished from other fibers such as those fibers used to make paper? The important definition of a textile fiber is that it must have an aspect ratio (i.e., the ratio of length to width) of at least 100. Most textile fibers have a much higher aspect ratio than this, but this minimum requirement at least enables them to be processed on regular textile carding, yarn spinning, and fabrication equipment. On the other hand, most paper fibers made from cellulose have an aspect ratio of less than 100, which means that they can only be processed into thin randomly distributed mats or sheets of paper using a water-based fiber suspension, from which the water is removed by suction, pressing, and drying processes.

In order to convert polymer resins into textile fibers or filaments, one needs to apply one of the following extrusion spinning techniques. The resulting fibers can either be continuous monofilament yarns or multifilament yarns or, alternatively, they can be extrusion spun, crimped, and cut into short-length staple fibers for blending with natural fibers such as cotton or wool. Usually, the units of filament size are expressed in terms of yarn linear density such as, decitex (dtex) (which is defined as the mass in grams of 10,000 m of yarn) or denier (more commonly used in North America and defined as the mass in grams of 9,000 m of yarn). The most commonly used extrusion spinning techniques are discussed below.

A) Melt Spinning As the name suggests, in melt spinning, the polymer resin is heated above its melting point and extruded through a spinneret containing one or more fine holes. Melt spinning is typically applied to thermoplastic polymers that are not affected by the high temperatures required in the melt spinning process. For a monofilament yarn, the spinneret consists of just one hole, while for a multi-filament yarn, the spinneret consists of a large number of finer holes resulting in finer individual filaments which are then entangled with an air jet or twisted to form a higher denier multifilament yarn. A schematic diagram of a typical melt spinning process is shown in Fig. 3.2. The dried polymer is heated in an extruder and then forced through a filter pack and the spinneret by a metering pump. The emerging molten thread line is then quenched in cold air to solidify the yarn. It is then drawn, lubricated, entangled, or twisted prior to winding onto a bobbin. The drawing operation helps to generate yarns with a finer diameter and improves their mechanical properties (e.g., tensile strength and modulus). Spin finish is applied as a lubricant to reduce yarn friction and improve yarn handling and efficiency during subsequent textile processes such as texturizing, weaving, and knitting. Depending on the shape of the spinneret hole, melt spun fibers can have various cross-sectional shapes such as round, trilobal, or hollow (Fig. 3.3).

B) Wet Spinning/Gel Spinning If the polymer is not thermoplastic or degrades when heated at elevated temperatures, a low-temperature wet spinning or gel spinning process is used. In the wet spinning process, the polymer is dissolved in a solvent and then extruded through a spinneret into a non-solvent in a coagulation bath where the fibers are precipitated and solidified. The filaments are then washed to remove all remaining solvents and non-solvents, drawn, dried, and lubricated, before winding up on a bobbin (Fig. 3.4). Viscose rayon, chitosan, and alginate fibers are produced by a wet spinning process.

Gel spinning differs from the wet spinning process because the polymer is actually not in a liquid solution during extrusion. The polymer is in the form of a gel with the polymer chains bound together at various points in a liquid crystalline structure. This results in filaments with strong interchain forces and a high degree of orientation which significantly increases the tensile properties. High-strength polyethylene and aramid fibers are produced by a gel spinning technique.

C) Dry Spinning Just like wet spinning, dry spinning also involves dissolving the polymer in a solvent. However, instead of precipitating the fibers in a coagulation

Fig. 3.2 Schematic of melt spinning process

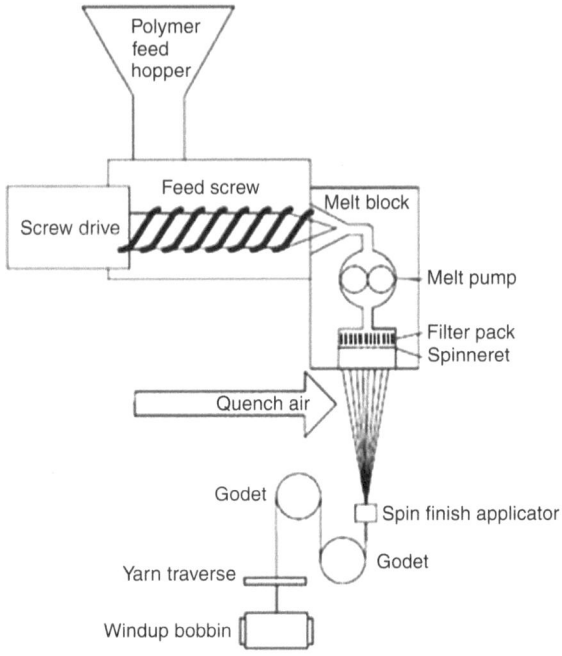

Fig. 3.3 Different fiber cross-sectional shapes: **a** Round nylon; **b** Trilobal nylon; **c** Round polyester with lower denier; **d** Crenulated viscose rayon [7]

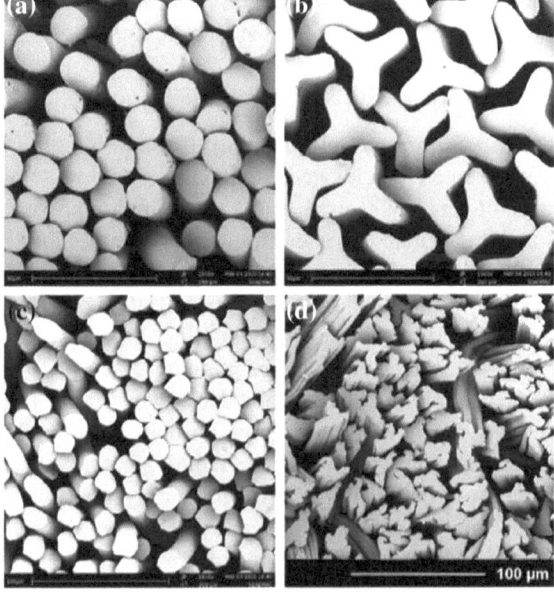

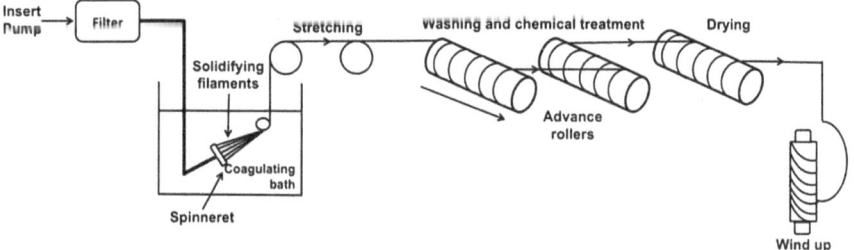

Fig. 3.4 Schematic diagram of the wet spinning process

Fig. 3.5 Schematic diagram
of dry spinning process

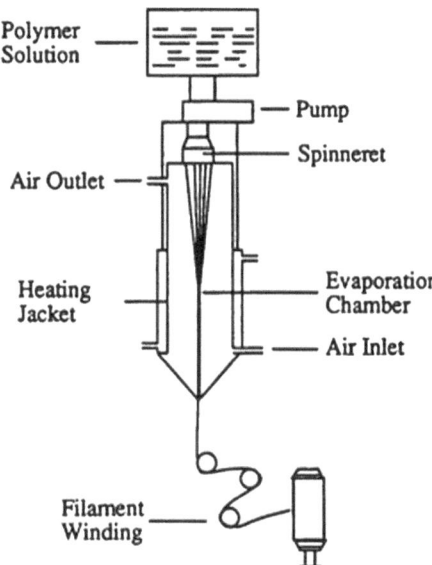

bath, dry spinning involves solidifying the fibers by evaporating the solvent in a stream of hot air or hot inert gas. The solvent is recovered and reused (Fig. 3.5). Acetate, triacetate, acrylic, modacrylic, and spandex fibers are typical examples of polymers that are normally dry spun.

D) Bicomponent Spinning Bicomponent, hybrid, or multicomponent fiber spinning is an advanced form of the melt spinning technique. This involves bringing two or more different streams of polymers together at the spinneret hole and co-extruding them so that the combined spun filament contains all the polymer components in different parts of the cross section. Figure 3.6 shows a number of bicomponent fiber cross-sectional configurations that have been developed.

   While theoretically one can propose a combination of any two or more thermoplastic polymers in order to spin a bicomponent fiber, in actual practice, the

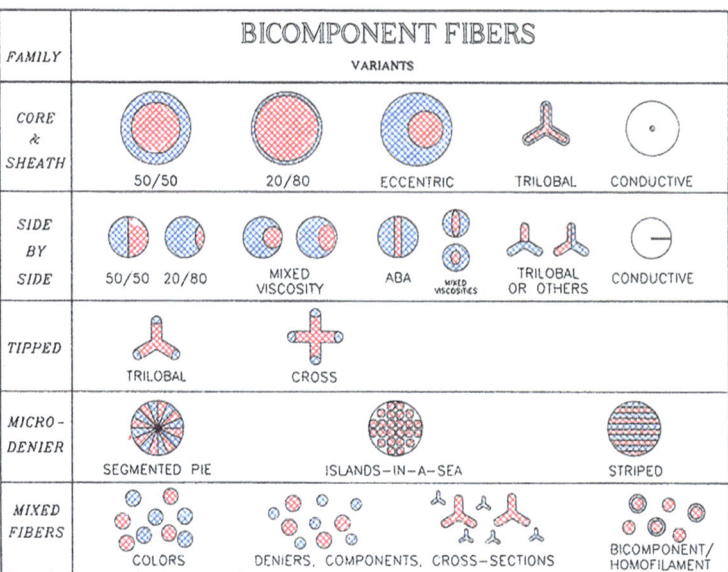

Fig. 3.6 Bicomponent fiber configurations [8]

selection of which two polymers to combine is somewhat limited. For example, it is necessary to combine polymers with similar melting temperature ranges, and it is important that the melt viscosities at their respective temperatures are also similar. Otherwise, the desired cross-sectional configuration and shape will either be distorted or the variants in Fig. 3.6 will not be realized.

The main motivations for developing such fibers are (1) to take advantage of the properties of more than one polymer component within the same fiber and (2) to achieve smaller fiber diameters by splitting or separating the fiber into its different components after co-spinning. For example, one of the configurations of a bicomponent fiber is to spin a resorbable polymer sheath around an inner core of a non-resorbable polymer. Here, the advantage is that the sheath modulates a faster inflammatory or foreign body response and a more complete healing process, while the core component maintains the mechanical integrity of the device. Additionally, drugs can be incorporated into the outer resorbable sheath and delivered at predefined rates, depending on the choice and thickness of the outer polymer.

Another approach is to make ultra-fine (submicron diameter) filaments by the island-in-the-sea technique, where the sacrificial matrix (sea) polymer is removed in a subsequent post-spinning process. Thus, by using bicomponent spinning technology, the material strength profile, resorption profile, and the biological properties can be engineered into the fiber to meet specific medical requirements.

3.3.2 Fabrication of Textile Structures

After a fiber or yarn is produced, it is then fabricated into a textile structure in order to obtain the desired form, shape, and mechanical properties for a medical device. There are four alternative types of textile structures that are typically used for medical devices. They include wovens, knits, braids, and nonwovens. Each structure has its own advantages and disadvantages. For example, woven fabrics are usually stronger and more dimensionally stable and can be fabricated with lower porosities, but are stiffer, less flexible, and more difficult to handle. Knits, on the other hand, have higher permeability and flexibility compared to woven fabrics, but may dilate after implantation. Braids have high longitudinal tensile properties, but can be unstable when subjected to torsional loads. Thus, the type of textile structure should be carefully selected when designing the biotextile device, and the medical application and the site of implantation should be taken into account.

A) Woven A woven fabric is fabricated by interlacing two sets of yarns perpendicular to each other on a loom or weaving machine. The yarns in the machine direction are called warp yarns and those in the cross-direction are weft or filling yarns. Woven fabrics have low elongation and high breaking strength in both directions and can be designed to have low porosity and water/blood permeability. Commonly used woven constructions for biotextile devices include plain, twill, satin, and leno weaves (Fig. 3.7) as well as combinations of these woven designs.

B) Knitted Knitted fabrics are made by interlooping yarns in horizontal rows (wales) and vertical columns (courses) of stitches. Generally, textured yarns are used for knitted structures so as to impart thickness and provide more softness and flexibility to the device. Knitted structures are characterized by having a large pore size, which allows tissue ingrowth. The total porosity of knitted structures is inevitably greater than 65 %, and thus, some applications require special processing to shrink the yarns and tighten the looped structure by heat setting. Weft knits and warp knits are the two most common constructions of knitted fabrics used for medical devices (Fig. 3.8). Warp-knitted structures have less stretch than weft knits, and therefore, they are more dimensionally stable and less prone to dilation in vivo.

A spacer fabric is an interesting knitted structure, which has a great potential as a tissue engineering scaffold. Spacer fabrics have two separate independently knitted fabric faces that are connected by a third distinct spacer yarn (Fig. 3.9). This forms a three-dimensional (3D) fabric structure having good breathability and resilience. The open pores and the 3D structure facilitate cell attachment, proliferation, and migration.

C) Braided Common braided structures are formed by the diagonal intersection of an even number of yarns that are interlaced at different angles and frequencies. The parameters that define the braided construction include the horizontal repeat distance l (called a line), the vertical repeat distance s (called a stitch), the width of

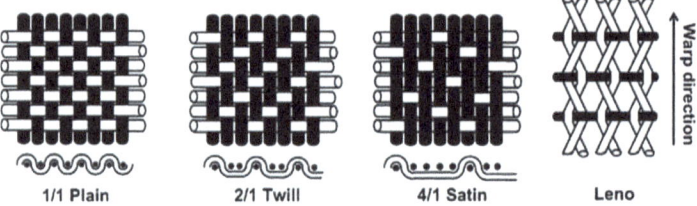

Fig. 3.7 Schematic diagram of typical woven designs

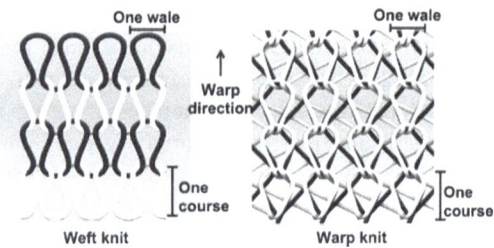

Fig. 3.8 Schematic diagram of knitted structures

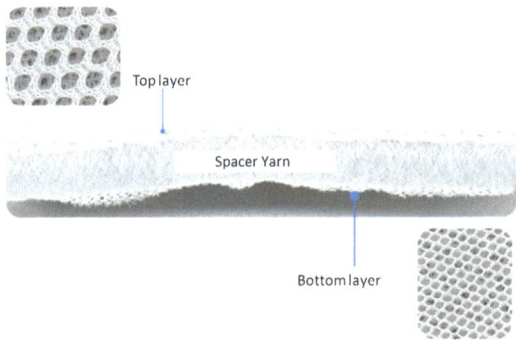

Fig. 3.9 Schematic diagram of 3D spacer fabric

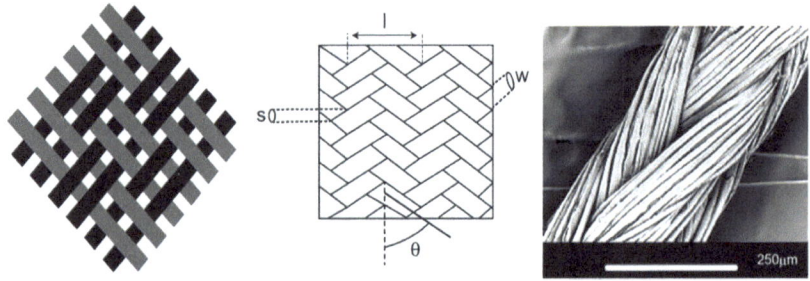

Fig. 3.10 Schematic diagram of braided structures (*left*) and SEM photomicrograph of a braided silk suture (*right*) [7]

the yarn *w*, and the braid angle θ between the yarn and the machine direction (Fig. 3.10). Currently, for medical applications, braided structures are used for sutures and anterior cruciate ligament (ACL) prostheses. They are also being studied for nerve guide conduits [9]. The challenge for using a braided structure as an implantable device is in securing the loose yarns at both ends of the device. Moreover, they are unstable when subjected to torsional loads.

D) Nonwoven As the name suggests, nonwoven fabrics are produced directly from fibers without the intermediate step of producing a yarn which is required for weaving. Short staple fibers or continuous filaments are laid down in a two-dimensional web and then bonded by mechanical, hydraulic, or thermal action, or by using an adhesive or solvent, or a combination of these bonding methods. Recently, techniques such as meltblowing and spunbonding have enabled the formation of nonwovens directly from polymer resins. The fibers may be oriented randomly or preferentially in one or more directions, and by combining multiple layers, like plywood, one can engineer the mechanical properties independently in the machine and cross-directions. The total porosity, average pore size, and pore size distribution of a nonwoven web can be controlled by changing certain variables, like the web thickness, fiber diameter and length, fiber orientation, and the method of bonding.

3.3.3 *Finishing*

Once the textile structure takes the size and shape of the final medical device, the next step is finishing. Finishing involves steps to clean the product in order to remove additives, lubricants, contaminants, and other impurities that might cause cytotoxicity or be associated with an unacceptably severe inflammatory response. Other finishing steps include heat setting, bleaching, shrinking, inspection, packaging, and sterilization. Since each polymer and fabrication process is different, the finishing operation varies and is material and device dependent. Special attention is required for the cleaning, packaging, and sterilization of devices made from resorbable polymers given their hydrolysable nature. This will be discussed in detail in a later section.

As can be seen from the process flow chart (Fig. 3.1), in order to fabricate and assemble a resorbable biotextile device, a resorbable polymer has to withstand exposure to a range of potentially 'hazardous' environments, such as thermal processing, mechanical abrasion, chemical treatment (lubricants, solvents, binders etc.), and atmospheric conditions, which include relative humidity. Each of these factors can drastically affect the rate of resorption and can cause 'premature degradation' of the resorbable implant.

References

1. M.W. King, Designing fabrics for blood vessel replacement. Can. Tex. J. **108**(4), 24–30 (1991)
2. H. Planck, M. Dauner, M. Renardy (eds.), *Medical Textiles for Implantation*. (Springer, Berlin, 1990)
3. S. Viju, G. Thilagavathi, B. Gupta, Preparation and properties of PLLA/PLCL fibres for potential use as a monofilament suture. J. Text. Inst. **101**, 835–841 (2010)
4. C.C. Chu, "Biodegradable Polymeric Biomaterials: An Updated Overview," in *Biomaterials: Principles and Applications*, ed. by J.B. Park, J.D. Bronzino, (CRC Press, Boca Raton, 2003), pp. 95–115
5. V.R. Gowariker, N.V. Viswanathan, J. Sreedhar, *Polymer Science*. (Wiley, New York, 1986)
6. R. Hill, E.E. Walker, Polymer constitution and fiber properties. J. Polym. Sci. **3**(5), 609–630 (1948)
7. M.W. King, S. Chung, "Medical Fibers and Biotextiles," in *Biomaterials Science: An Introduction to Materials in Medicine*, 3rd edn., ed by B.D. Ratner, A.S. Hoffman, F.J. Schoen, J.E. Lemons, (Academic Press, New York, 2012), pp. 301–320
8. Hills Inc., "An Introduction to Bicomponent Fibers." [Online]. Available: http://www.hillsinc.net/articles/bicointro.htm. [Accessed: 17-Dec-2013]
9. S. Ichihara, Y. Inada, T. Nakamura, Artificial nerve tubes and their application for repair of peripheral nerve injury: an update of current concepts. Injury **39**(Supplement 4), 29–39 (2008)

Chapter 4
Hydrolytically Sensitive Fiber-Forming Bioresorbable Polymers

Keywords Biopolymers · Linear aliphatic polyesters · Bacillus megaterium · Bacterial polyesters · PDS · Biopol · Maxon · Vicryl · Monocryl · Panacryl · Purasorb · Tepha

There are many different bioresorbable polymer systems based on different degradation mechanisms and having a range of physical and mechanical properties. However, the scope of our review will be restricted to only fiber-forming polymers. Those that are hydrolytically sensitive are discussed in this chapter, while enzymatically catalyzed bioresorbable polymers are presented in Chap. 6. Table 4.1 shows the classification of fiber-forming hydrolytically sensitive bioresorbable polymers.

4.1 Linear Aliphatic Polyesters

4.1.1 Poly(α-esters)

Poly(α-esters) are the earliest, most extensively studied, and commercially widely used bioresorbable polymers. Especially, the poly(α-hydroxy acids), including poly(glycolic acid) (PGA), stereoisomers of poly(lactic acid) (PLA) and their copolymers which have been extensively investigated for their degradation mechanisms and structure–property relationships [1–4]. Aliphatic polyesters based on poly(α-ester)s are discussed below.

4.1.1.1 Poly(glycolic acid)

Poly(glycolic acid) (PGA) is one of the first bioresorbable polymers investigated for biomedical applications. Dexon®, the first biodegradable synthetic suture to be approved by the Food and Drug Administration (FDA) in 1969, was based on

C. R. Gajjar and M. W. King, *Resorbable Fiber-Forming Polymers for Biotextile Applications*, SpringerBriefs in Materials, DOI: 10.1007/978-3-319-08305-6_4, © The Author(s) 2014

Table 4.1 Classification of fiber-forming hydrolytically sensitive bioresorbable synthetic polymers

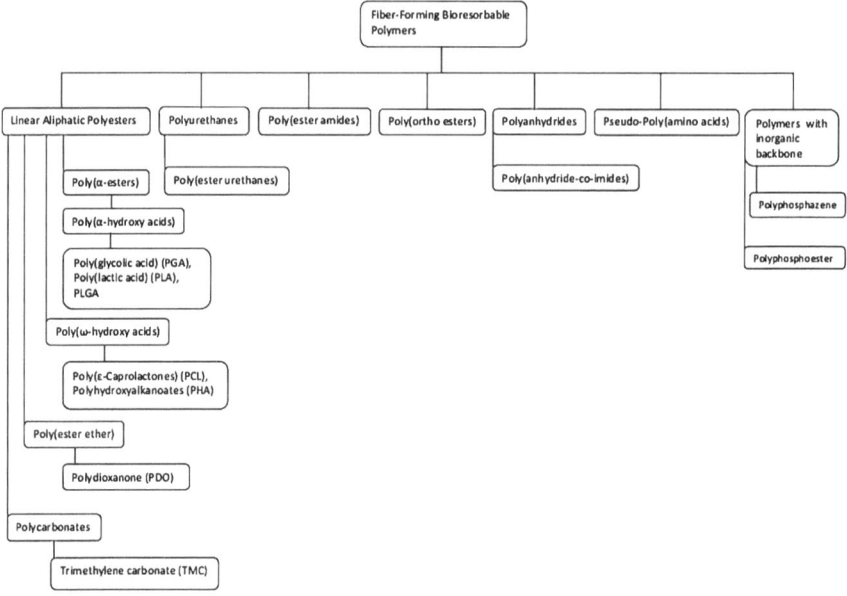

PGA. PGA is a highly crystalline polymer (45–55 % crystallinity), and thus, it has superior mechanical properties. For the same reason, its solubility is limited to only a few organic solvents such as hexafluoroisopropanol. The glass transition temperature (Tg) ranges from 36 to 40 °C, and the melting point (Tm) is 224–230 °C. It is a good fiber-forming polymer. Fibers from PGA show high tensile strength and modulus.

PGA is a bulk-degrading polymer. Its strength loss occurs in 1–2 months, and its total mass loss takes 4–12 months. Figure 4.1 shows the loss in tensile strength over time for PGA during in vitro degradation [5, 6]. In the body, polyglycolides are broken down to glycine which can be excreted in the urine or converted into carbon dioxide and water through the citric acid cycle [7]. Due to its rapid degradation, PGA has been studied for short-term tissue engineering scaffolds. It is often fabricated into a mesh network and used as a scaffold for bone, cartilage, tendon, dental, and spinal nerve regeneration [8].

Because of its rapid degradation rate, the formation of acidic degradation by-products, and its limited solubility, there are few opportunities to use of PGA as a homopolymer. It is therefore generally used as a copolymer with other resorbable biomaterials (Table 4.2).

Fig. 4.1 In vitro degradation of PGA—retained tensile strength versus time [6]

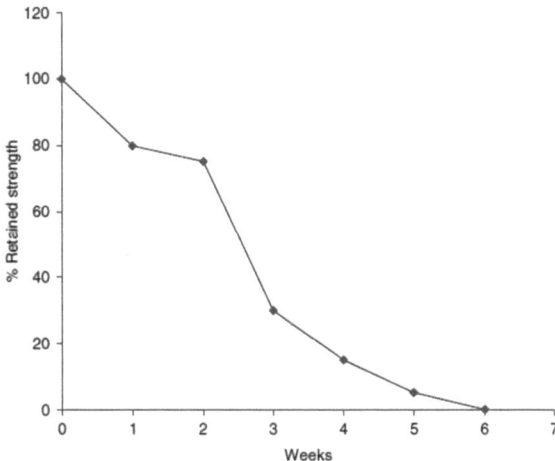

Table 4.2 PGA properties

PGA properties	
Melting temperature (Tm)	224–230 °C
Glass transition temperature (Tg)	36–40 °C
Crystallinity	45–55 %
Time for strength loss	1–2 months
Time for mass loss	4–12 months

4.1.1.2 Poly(lactic acid)

Poly(lactic acid) (PLA) is considered to be a well established bioresorbable polymer in terms of research as well as commercial aspects, due to its inherent biocompatibility, high mechanical strength and modulus, ease of processing, and availability from renewable natural sources such as corn. Unlike PGA, PLA is soluble in a wider range of organic solvents. PLA can form high strength, high modulus fibers and was approved by the FDA for suture applications in 1971. Since then, it has been studied extensively and used for a number of biomedical applications.

Because of the chirality of the polylactide molecule, PLA has four stereoisomeric forms: poly(L-lactic acid) (PLLA), poly(D-lactic acid) (PDLA), meso-poly(lactic acid), and poly(D,L-lactic acid) (PDLLA), which is a racemic mixture of PLLA and PDLA. L-lactide is a naturally occurring isomer. For the use in biomedical applications, only PLLA and PDLLA have been found to be promising biomaterials [8].

PLLA is a semicrystalline polymer (about 37 % crystalline) with high melting point (Tm) of 170–180 °C. Its Tg is 55–65 °C. PLLA exhibits high tensile strength (59 MPa) and low elongation (4–7 %) and consequently has a high modulus (3.8 GPa), which makes it more suitable for load-bearing applications such as in orthopedic fixation and surgical sutures [9, 10] (Table 4.3). Due to the high strength of PLLA fibers, it has been proposed for use as a scaffold material for ligament regeneration [11, 12]. PLLA fiber-based devices have also been evaluated as long-term blood vessel conduits [13].

However, the degradation rate of PLLA is very slow. High molecular weight PLLA can take between two and five and a half years for complete resorption [14, 15]. Even though the polymer appears to lose its strength in 6–12 months when hydrolyzed, no significant loss of mass occurs for an extended period. It is believed that the high Tg is mainly responsible for the extremely slow degradation rate [16]. In addition, the presence of methyl($-CH_3$) side groups impart a hydrophobic nature to PLLA, making the hydrolysis slower. The methyl substituent also causes a steric-shielding effect of the ester group, which further reduces the rate of hydrolytic degradation [9, 17]. Because of such a low rate of degradation, biomedical applications of PLLA and its copolymers with a high PLLA content have been restricted to orthopedic surgery, drug release devices, coating material for sutures, vascular grafts, and surgical meshes to facilitate healing after dental extraction [16].

PDLLA is an amorphous, transparent material with slightly lower Tg and lower strength. It is preferred for making low strength scaffolds for tissue engineering. Even so, PDLLA takes over a year for complete resorption.

To reduce the degradation time, PLA has been blended with several other degradable polymers. Resomer® LR708 is a 70:30 amorphous copolymer of poly(L-lactide-co-DL-lactide) and is being investigated as a biocompatible, bioresorbable implantable material [18]. Blends with other bioresorbable polymers such as poly(ethylene glycol) (PEG) and poly(ethylene oxide) (PEO) have also been explored. The most studied and commercially successful blend of PLA has been with PGA Poly(lactide-co-glycolide) (PLGA).

PLA degrades through a bulk erosion mechanism. Random chain scission of the ester linkage in the polymer backbone proceeds homogenously throughout the material. It has been reported that devices greater than 1 mm in thickness degrade via a heterogeneous mechanism, that is, the degradation proceeds more rapidly in the center than at the surface. This is attributed to the autocatalytic action of acidic degradation products trapped in the matrix [19]. Figure 4.2 shows the rate of in vitro degradation and loss in tensile strength over time for PLA [6].

4.1.1.3 Poly(lactide-co-glycolide)

PLGA is a versatile biopolymer since it allows the modulation of properties such as strength and rate of resorption. PLGA copolymers have a wide range of properties and applications, depending on the composition ratio of glycolide to lactide. The full range of PLGA copolymers has been explored for different applications. Both L- and DL-lactides have been used for copolymerization. PLGA copolymers are amorphous and show Tg in the range 40–60 °C. PLGA can be dissolved in a wide range of solvents such as acetone, ethyl acetate, tetrahydrofuran, and chlorinated solvents.

PLGA does not show a linear relationship between the copolymer composition and the mechanical and degradation properties, instead there is a typical U-shape

Fig. 4.2 In vitro degradation of PLA—retained tensile strength versus time [6]

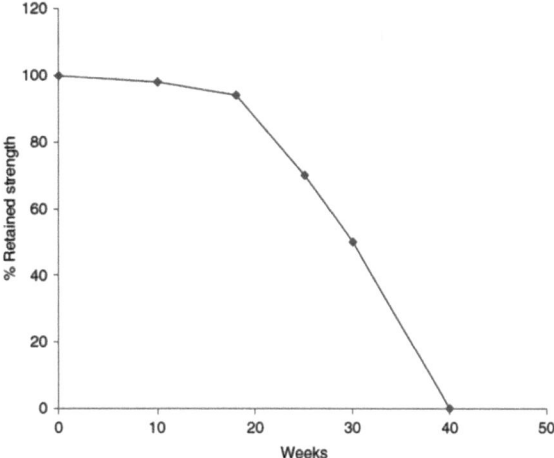

Table 4.3 PLLA properties

PLLA properties	
Melting temperature (Tm)	170–180 °C
Glass transition temperature (Tg)	55–65 °C
Crystallinity	37 %
Time for strength loss	6–12 months
Time for mass loss	24–66 months
Tensile strength	59 MPa
Tensile modulus	3.8 GPa
Elongation to break	4–7 %

relationship (Fig. 4.3). All the intermediate copolymers are less stable and hydrolyze faster than each of the homopolymers. Thus, the resistance to hydrolysis is more pronounced at either end of the copolymer composition range [20, 21]. 50/50 PLGA has been shown to be the most readily hydrolysable. 50/50 PLGA is resorbed completely in 1–2 months, 75L/25G in 4–5 months, and 85L/15G in 5–6 months [14].

PLGA has been used as a suture material for a long time. High concentrations of glycolide monomer are required to ensure adequate mechanical properties and a suitable rate of resorption for use as a suture. Vicryl® (Ethicon) is a multifilament Syneture braided suture introduced in 1974. It is a copolymer containing 90 % glycolic acid and 10 % lactic acid. It has a longer strength-retention time and takes 3–4 months for complete resorption. Vicryl Rapid® is the newer, irradiated version of the suture with a shorter time for resorption. Vicryl Mesh® is used as the scaffold for the tissue engineered skin graft Dermagraft®. Purasorb® (Purac Bio-materials), Panacryl® (Ethicon), and Polysorb® (Syneture) are other suture materials composed of PLGA. Purasorb® is a copolymer with the ratio 80L:20G, while Panacryl® has a ratio of 90L:10G. Both materials have more rapid resorption

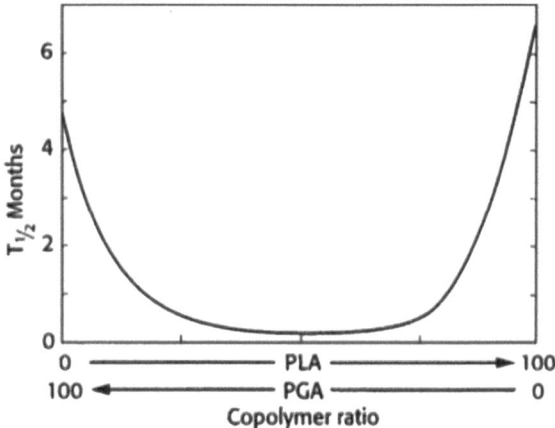

Fig. 4.3 Half-life for PLGA copolymers with different copolymer ratios, implanted in rat tissue [14]

rates than Vicryl®. PLGA is an excellent biomaterial for tissue engineering scaffolds as well. It shows good cell adhesion and proliferation properties. PLGA has also been used for guided tissue regeneration (Cytoplast Resorb®).

Just like the homopolymers, PLGA degrades by a bulk erosion mechanism. Figure 4.3 shows the in vivo half-life degradation rate for PLGA copolymers with different copolymer ratios when implanted in rat tissue [14].

4.1.2 Poly(ω-hydroxy acids)

4.1.2.1 Poly(ε-caprolactone)

Polycaprolactone (PCL) is a semicrystalline polymer based on poly(ω-hydroxy acids). It is obtained from ε-caprolactone. PCL has a Tg of −60 °C and a Tm of 55–60 °C. It is soluble in a wide range of organic solvents and forms miscible blends with various polymers. PCL has low tensile strength (23 MPa), but high elongation at break (700 %) [8]. Thus, it is a highly elastic biomaterial. However, the degradation rate of PCL is very slow, normally taking 2–3 years. Hence, PCL is often blended or copolymerized with other biopolymers such as PLLA [22, 23], PDLLA, PLGA [24], and polyethers. For example, a copolymer of glycolide and ε-caprolactone is commercially available as a monofilament suture called Monacryl®. It is less stiff than PGA and shows superior handling properties, minimal tissue drag, and good tensile properties. The resorption rate of Monacryl® is faster than that of PDS® monofilament sutures [25, 26]. A copolymer of L-lactide and ε-caprolactone also shows satisfactory strength and flexibility, which is appropriate for a monofilament suture material. Despite the low straight-pull tensile strength, it shows good knot security [27]. Another bioresorbable multiblock copolymer composed of caprolactone, glycolide, lactide, and PEG units has been

Table 4.4 PCL properties

PCL properties	
Melting temperature (Tm)	55–60 °C
Glass transition temperature (Tg)	−60 °C
Crystallinity	Semicrystalline
Time for mass loss	24–36 months
Tensile strength	23 MPa
Elongation to break	700 %

developed for drug delivery applications. It is marketed as SynBiosys®. So, PCL and its copolymers are being widely used for drug delivery applications and as tissue engineering scaffolds (Table 4.4).

4.1.2.2 Polyhydroxyalkanoates

Polyhydroxyalkanoates (PHAs), also known as 'bacterial polyesters', are linear aliphatic polyesters based on poly(ω-hydroxy acids). They are synthesized by bacteria as storage compounds for energy and carbon. PHAs exhibit a high degree of polymerization and command high molecular weights with values up to several million Daltons [28]. They are biocompatible, bioresorbable, and piezoelectric thermoplastic polymers.

Poly(3-hydroxybutyric acid) (P(3HB)) is the most common polymer in this class. It was first synthesized from the bacterium *Bacillus megaterium* in 1926. Later, other useful hydroxyalkanoates such as poly(4-hydroxybutyric acid) (P(4HB)) and poly(3-hydroxyvalerate) (P(3HV)) were discovered. PHAs show a wide range of mechanical and resorption properties. P(3HB) is a stiff, rigid, brittle polymer having a tensile strength of about 40 MPa. It has a 6 % elongation at break, and its tensile modulus is 3.5 GPa. P(3HB) has a Tm in the range of 160–180 °C, and its Tg is around 5 °C. On the other hand, P(4HB) is a highly ductile and flexible polymer with an elongation at break in the range of 1,000 %. Its tensile strength is 104 MPa, and modulus is 0.149 GPa. It has a low Tg at −51 °C, and its Tm is 60 °C. It is soluble in a wide range of solvents like acetone, and it is fairly stable in the melt up to about 180 °C. The rate of bioresorption of PHAs also varies widely and depends mainly on their chemical composition. Other determining factors include surface area, physical shape, degree of crystallinity, and molecular weight [29].

PHAs have been explored for many different biomedical applications [30]. P(3HB) and P(4HB) fibers have been evaluated in animal models for suture applications [31]. Nonwoven webs of P(3HB) and P(4HB) fibers have been studied for use in nerve repair conduits [32, 33]. P(3HB) fibers have been investigated as the carrier scaffold to promote axonal migration and help spinal cord regeneration [34]. Due to its piezoelectric activity, P(3HB) is a promising candidate as a scaffold material for bone regeneration. Some studies have indicated a favorable osteogenic response to P(3HB) [35, 36]. P(4HB) fibers and multifilament yarns

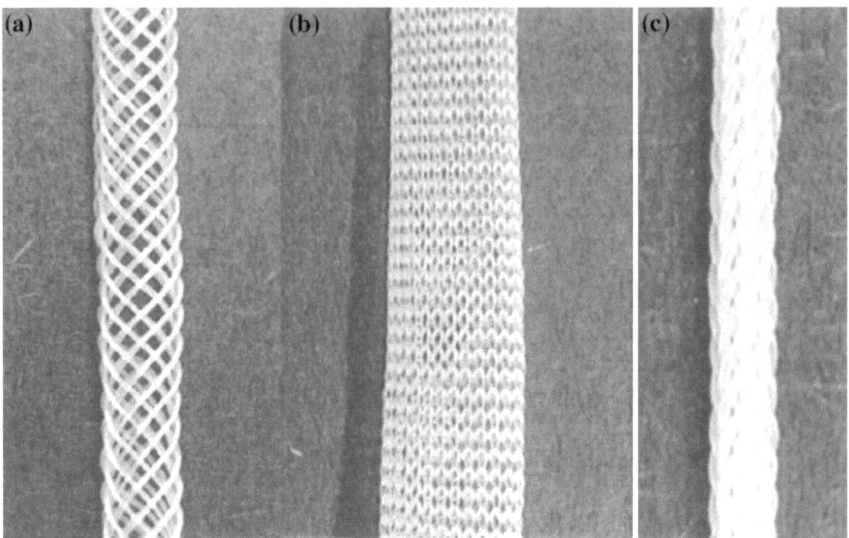

Fig. 4.4 Textile structures fabricated from P(4HB): **a** 16 filament-braided tube; **b** circular weft-knitted tube, **c** 16 filament braid [37]

with their wide range of mechanical and resorption properties have been proposed for biotextile products such as grafts, patches, scaffolds, ligaments, prolapse slings, hernia meshes as well as dura mater and pericardial substitutes (Fig. 4.4) [37]. P(4HB) sutures, with a tensile strength in the range of 410–460 MPa, have been developed. This is comparable to commercial absorbable sutures such as PDS II® (450–560 MPa) and Maxon® (540–610 MPa). In addition, P(4HB) sutures have the unique advantage of a low Young's modulus compared with other fibers. The elastic modulus for P(4HB) sutures is only 670 MPa compared with 2,930 MPa for Maxon® and 1,380 MPa for PDS II®. These mechanical properties lead to improved handling and a better breaking strength-retention profile.

Since the homopolymer P(3HB) is a tough and brittle polymer, it has been copolymerized with P(3HV). The copolymer P(3HB-co-3HV) has greater potential as a biomaterial. Its Tg is between −5 and 20 °C, and its Tm is 80–160 °C depending upon the composition. It has been studied for use as a suture material, an orthopedic implant as well as an adhesion prevention film. It was first made commercially available by Imperial Chemical Industries (ICI) under the trade name Biopol®, but now the product has been discontinued. Currently, Tepha™ (a spinoff from Metabolix Inc.) is actively engaged in developing biotextile products based on PHAs.

PHAs degrade by a surface erosion mechanism. Hence, there is a gradual evolution of the mechanical properties rather than an abrupt change as observed with PGA and PLA. PHAs degrade more slowly than PGA, but faster than PLLA and PCL. P4(HB) resorbs in 2–12 months, while P(3HB) takes about 2 years or more. The copolymer P(3HB-co-3HV) degrades at a much faster rate. However, to

Table 4.5 PHA properties

P(3HB) properties	
Melting temperature (Tm)	160–180°C
Glass transition temperature (Tg)	5 °C
Time for mass loss	24–30 months
Tensile strength	40 MPa
Tensile modulus	3.5 GPa
Elongation to break	6 %
P(4HB) properties	
Melting temperature (Tm)	60 °C; stable in the melt up to 180 °C
Glass transition temperature (Tg)	−51 °C
Time for mass loss	2–12 months
Tensile strength	104 MPa
Tensile modulus	0.149 GPa
Elongation to break	1,000 %

date no correlation has been found between the degradation rate and the proportion of HV in the copolymer. Hydrolytic degradation of P(3HB) results in the formation of D-(−)-3-hydroxy-butyric acid, which is a normal constituent of blood.

PHAs have several advantages over conventional resorbable polymers such as PLA and PGA. They offer superior mechanical properties and a less severe inflammatory response in vivo owing to the production of less acidic by-products. They are generally less sensitive to hydrolysis by atmospheric or residual moisture and thus have longer shelf stability [37]. At present, commercialization of PHAs is impeded by the high cost of production. Nevertheless, PHAs have recently become one of the leading classes of biomaterials under investigation for tissue engineering scaffolds (Table 4.5).

4.1.3 Poly(ester ethers)

4.1.3.1 Polydioxanone

Fiber-based applications of PLA and PGA such as sutures require the use of fine multifilament braided structures in order to attain the necessary flexibility. However, the disadvantages are that the multifilament yarns are associated with greater polymer-tissue friction and higher rates of infection. As a result, the more flexible polymer, polydioxanone (PDO) was developed as a thicker monofilament suture. PDO is a glycolide-derived bioresorbable polymer having poly(ester ether) linkages, which impart greater flexibility. Incorporation of the ether segment into the repeat unit reduces the density of ester linkages and reduces intermolecular hydrogen bonds and thus improves the flexibility. However, fewer ester linkages results in a slower degradation rate [16].

Fig. 4.5 In vitro strength
loss and mass loss of 2-0
PDS® sutures as a function of
hydrolysis time [39]

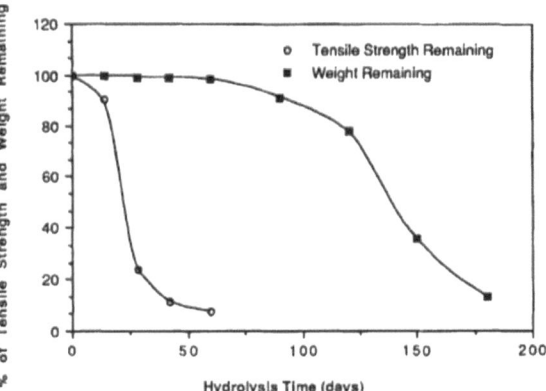

PDO is a semicrystalline polymer with a low Tg (-10–0 °C) and Tm in the range of 106–115 °C. PDO fibers have a low tensile modulus (1.5 GPa) and good flexibility. PDS® was the first monofilament suture based on PDO that was introduced in the 1980s. PDS® sutures have sufficient flexibility and ease of handling that they are acceptable for use in vascular applications. They have good biocompatibility and superior tensile strength and knot security. However, the degradation rate of PDO is not very fast. The high crystallinity (around 55 %) and hydrophobicity result in a moderate rate of resorption. PDO loses its strength within 1–2 months and its mass within 6–12 months [7, 38]. Moreover, there is a large reduction in tensile strength before there is any significant mass loss. For example, when PDO sutures were incubated at 37 °C and pH 7.4 in a phosphate buffer solution (PBS) for 60 days, more than 90 % of the initial tensile strength was lost while only 1.5 % of the mass was lost (Fig. 4.5) [39]. An improved version, PDS II®, was introduced with an extended tensile strength loss profile. This was achieved by recrystallizing the melt spun fibers by heat setting them at 128 °C for a short period of time [16].

Various copolymers with a high PDO content have been synthesized and studied to improve the mechanical performance and increase the rate of resorption [40]. A copolymer with 80 % PDO and up to 20 % PGA has a resorption profile similar to Dexon® and Vicryl® and yet maintains a compliance similar to PDS®. The copolymer with 85 % PDO and up to 15 % PLLA exhibits higher compliance and a lower modulus than the PDO homopolymer, yet its resorption profile is similar to PDS® [41]. Copolymers with three or more different monomer units have also been explored. One such example is poly(GA-co-PDO-co-LA), which has been proposed as a possible suture material with suitable crystallinity, good flexibility, better strength retention, and a reasonably rapid resorption rate [42]. Until recently, there was not much interest in developing PDO as a biomaterial, mainly because of the lack of commercial availability and difficulties in its synthesis. However, with the development of new alternative synthesis procedures and the availability of the PDO monomer, there has been more commercial interest in developing PDO as a bioresorbable polymer in recent years (Table 4.6).

Table 4.6 PDO properties

PDO properties	
Melting temperature (Tm)	106–115 °C
Glass transition temperature (Tg)	−10–0 °C
Crystallinity	55 %
Time for strength loss	1–2 months
Time for mass loss	6–12 months
Tensile modulus	1.5 GPa

4.1.4 Polycarbonates

4.1.4.1 Poly(trimethylene carbonate)

Poly(trimethylene carbonate) (PTMC) is the most extensively studied and commonly used polycarbonate. It is an elastomeric polymer with Tg of −17 °C. While it has good flexibility and a slow degradation profile, it has inadequate mechanical strength. The inferior mechanical performance of the PTMC homopolymer limits its applications, and as a result, several copolymers have been developed with PLA, PGA, and other resorbable biopolymers. These copolymers have superior mechanical properties and a predictable degradation profile. Block copolymers of glycolides and TMC (polyglyconates) have been developed for use as flexible suture materials, one of which is commercially available as Maxon®. Polyglyconates are generally prepared as A-B-A block copolymers in 2:1 GL/TMC ratio, with a GL-TMC center block (B) and pure GL end blocks (A). The resulting copolymer has superior flexibility compared with pure PGA. Maxon® sutures have a lower coefficient of friction, lower tissue drag, superior tensile strength and knot security than corresponding sizes of braided synthetic absorbable sutures. Maxon® loses 50 % of its initial strength in 5 weeks and the mass resorbs completely in 6–7 months. BioSyn® is a terpolymer composed of glycolide, TMC and PDO that has reduced stiffness and degrades within 3–4 months. It was developed for use as a suture material. Polycarbonates degrade by a surface erosion mechanism. The major advantage of incorporating TMC units in the copolymer is that the degradation by-products derived from TMC are neutral in pH.

4.2 Polyurethanes

Thermoplastic polyurethanes (TPUs) are synthesized from three monomers, namely a diisocyanate, a diol or diamine chain extender, and a long-chain diol. These monomers react to form linear, segmented copolymer chains consisting of alternating hard and soft segments. The hard segment consists of the alternating diisocyanate and chain extender molecule, while the soft segment consists of the long-chain linear diol. Due to their excellent biocompatibility, mechanical properties, and structural versatility, polyurethanes have been widely used for long-term medical implants such as cardiac pacemakers and vascular grafts. However, despite their good biological performance, the long-term biostability of polyurethanes has remained a limiting question. As a result, attempts have been made to develop bioresorbable polyurethanes.

In designing bioresorbable polyurethanes, the chemical structure of the diisocyanate and the diol plays an important role. With the appropriate selection of these compounds and their relative proportions, polyurethanes can be synthesized with a wide range of mechanical properties and bioresorption profiles. Diisocyanates are toxic compounds. However, the resulting polyurethane after hydrolytic degradation does not release the diisocyanate, but the corresponding diamine. Thus, the choice of the diisocyanate is governed by the toxicity of the corresponding diamine. Biocompatible aliphatic diisocyanates such as ethyl lysine diisocyanate (ELDI), methyl lysine diisocyanate (MLDI), hexamethylene diisocyanate (HDI), and 1,4-butanediisocyanate (BDI) are some of the more commonly used diisocyanates. Diols based on polyesters are preferred for the soft segment. Among all these systems, HDI-based bioresorbable polyurethanes are the most widely studied. They have high tensile strength (60 MPa) and elongation at break of 580 %. HDI-based polyurethanes can form fibers with both superior strength and flexibility [43, 44].

Bioresorbable poly(ester urethanes) have been developed by reacting lysine diisocyanate (LDI) with polyester diols or triols based on D,L-lactide, caprolactone, and other copolymers [45]. In these systems, aliphatic polyesters such as PLGA or PCL form the soft segments and the polypeptides form the hard segments [46]. A resorbable elastomeric poly(ester urethane) called Degrapol® is available commercially. It is currently being used to develop porous scaffolds for tissue engineering applications.

4.3 Poly(ester amides)

Polyesteramides contain both amide as well as ester linkages, which impart amphiphilicity and bioresorbability. They were designed to combine the excellent mechanical properties of polyamides and the resorbability of polyesters [47]. Because of the polar nature of amide groups and their ability to form hydrogen bonds, these polymers exhibit good thermal and mechanical properties even at low molecular weights. In addition, the hydrolytically degradable ester bond provides bioresorbability to the polymer. Another advantage of these polymers is their ability to incorporate α-amino acids, which provide sites for enzymatic degradation [44].

Owing to their superior mechanical properties derived from the symmetrical bisamide-diols and succinyl chloride groups, poly(ester amides) have been investigated as potential candidates for bioresorbable suture materials [18]. Attempts have also been made to increase the degradation rate of poly(ester amides) by incorporating amino acid units into the polymer backbone. Cameo® is an example of a commercial poly(ester amide) that has been developed for drug delivery applications.

4.4 Poly(ortho esters)

Poly(ortho esters) (POEs) are hydrophobic, surface-eroding polymers whose performance properties, such as their rate of resorption, Tg, and pH sensitivity, can be varied by selecting different types of diols with varying levels of chain flexibility. Four different classes of POEs with different mechanical properties and degradation rates have been developed [48–51]. POE I is synthesized by the transesterification between a diol and diethoxytetrahydrofuran. The acidic by-product has an autocatalytic effect on the degradation rate of the polymer. POE II has been designed to overcome the autocatalytic effect of POE I. It is synthesized by reacting a diol with a diketene acetal. The degradation by-products are molecules with a neutral pH. The degradation rate of this polymer can be modulated by incorporating dibasic acids such as itaconic or adipic acid. POE III is synthesized by the direct polymerization of a triol with an ortho ester. In this case, the polymer chains are highly flexible causing the polymer to behave like a gel at room temperature, which makes it a suitable biomaterial for drug delivery applications. However, the technical difficulties in scaling up the polymerization procedure, and the gel-like consistency of the polymer limit the commercial applications of POE III. This has led to the development of POE IV, which is synthesized by a similar procedure to that for POE II without the addition of an acid excipient. This is achieved by incorporating short segments of lactic or glycolic acid into the polymer backbone. The rate of resorption for these polymers can then be modulated from a few days to several months by varying the amount of the acid segment in the polymer backbone [52]. Moreover, by altering the size and structure of the diol, the nature of the polymer obtained can either be a hard solid or a soft gel-like material.

On account of their controlled resorption rate and biocompatibility, POEs have been explored mainly as drug delivery vehicles. Their use as tissue engineering scaffolds is limited due to their weak mechanical properties. There are no reports in the literature for their use as resorbable suture materials.

4.5 Polyanhydrides

Aliphatic polyanhydrides were developed in 1932 as a fiber-forming polymer for textile applications. However, the hydrolytic instability of the polymer limited its use as a textile fiber. More recently, the surface erosion phenomenon of the polymer responsible for its hydrolytic instability has made it attractive as a resorbable biomaterial. Polyanhydrides are one of the most extensively investigated surface-eroding biomaterials for applications related to controlled drug release [53–55]. In fact they are FDA approved as drug delivery materials. Many homo- and co-polyanhydrides have been developed with different properties. Aliphatic homo-polyanhydrides, such as poly(sebacic anhydride), have limited applications due to their inferior mechanical properties and rapid rate of resorption. As a result several copolymers have been developed. For example, copolymers of sebacic anhydride and hydrophobic aromatic comonomers have been studied with the objective of developing polymeric systems with controllable rates of resorption and ease of processing. In order to develop surface-eroding polymers with good mechanical strength, poly(anhydrides-co-imides) have been developed [56, 57]. The imide segments in the polymer backbone imparts high strength. Recently, poly(anhydride ester)s have been explored for biomedical applications. These polymers have a Tg lower than physiological temperatures (12–34 °C) and are often copolymerized with traditional aromatic diacid monomers to improve their mechanical properties. Salicylic acid-based poly(anhydride ester)s and their composites have been fabricated into particles, films, and fibers for a range of different biomedical applications [58–60].

4.6 Pseudo-poly(amino acids)

Poly(amino acids) are naturally occurring biopolymers. However, their application as a biomaterial has been limited due to their immunogenic response and their poor mechanical performance. To overcome these limitations, attempts have been made to develop pseudo-amino acids in which amino acid like backbones are connected by carbonate, ester, or imino carbonate bonds giving them strong mechanical properties while maintaining the biocompatibility of their degradation by-products. One of the most commonly studied pseudo-poly(amino acids) is the family of tyrosine-derived polycarbonates, namely the poly(desaminotyrosyl-tyrosine alkyl ester carbonates) (PDTEs). Because of the aromatic groups in the polymer backbone, PDTEs have sufficiently high mechanical strength required for load-bearing applications. They have a pendant alkyl chain of variable length allowing for modulation of their thermal and mechanical properties [61]. The Tg ranges from 50 to 81 °C, and Tm ranges from 75 to 118 °C. PDTEs have a tensile strength in the range of 50–70 MPa and elastic modulus or stiffness of 1–2 GPa. One significant difference between PLLA and tyrosine-derived carbonates is in their ability to absorb water. The tyrosine-derived polymers do not absorb more than 5 % water even at the later stages of degradation and are therefore able to maintain their shape for a longer period of time. Another unique advantage is that the polymer experiences a mass loss only at the very end of the degradation process [18]. The low acidity of the degradation by-products is another advantage [62, 63]. Slow resorption over 200 days for the loss of 50 % of its original strength and minimal mass loss allows PDTEs to maintain their physical properties for a longer

Table 4.7 Pseudo-
poly(amino acid) properties

Pseudo-poly(amino acid) properties	
Melting temperature (Tm)	75–118 °C
Glass transition temperature (Tg)	50–81 °C
Time for mass loss	18–30 months
Tensile strength	50–70 MPa
Tensile modulus	1–2 GPa

time making them suitable candidates for applications involving slow tissue regeneration [64].

Good processability has facilitated the fabrication of fibers and films out of PDTEs [8, 65]. They form high strength fibers by melt extrusion with tensile strengths of about 230 MPa and Young's modulus in the range of 3.1 GPa. The fibers have been shown to retain 87 % of the strength after 30 weeks of incubation in water [66]. The tyrosine-derived polyarylates generate softer elastomeric forms of pseudo(amino acid) polymers having faster resorption rates than the polycarbonates (Table 4.7).

4.7 Polyphosphazenes

Polyphosphazenes are a unique class of resorbable polymers in that their backbone is completely inorganic consisting of phosphorous and nitrogen bonded linearly through alternating single and double bonds [18]. These polymer chains are highly flexible both physically and chemically [67]. The selection of the two side groups attached to the phosphorous influences such properties as the biocompatibility, flexibility, dipole moment, Tg, chemical inertness, elastomeric performance, and mechanical strength. Using different substituent side groups, thermal and mechanical properties have been shown to vary greatly with Tg varying from −10 to 35 °C, tensile strength from 2.4 to 7.6 MPa, and modulus of elasticity ranging from 31.4 to 455.9 MPa [68]. Thus, the size and chemistry of the attached side groups allows for the design of biomaterials with highly controlled structure and properties such as crystallinity, solubility, hydrophilicity, thermal transitions and desired resorption profiles ranging from a few hours to several years.

Polyphosphazenes have shown promising results for drug delivery and tissue engineering applications. Scaffolds for nerve regeneration and orthopedic applications have been fabricated using polyphosphazene films, fibers, and sintered microspheres [69–72]. Poly[(amino acid ester) phosphazenes] have shown the most potential for biomedical applications since they undergo both surface and bulk erosion, and form neutral by-products on degradation. Polyphosphazenes have also been blended with polyesters to increase their mechanical strength.

4.8 Polyphosphoesters

Polyphosphoesters are biomaterials composed of phosphorus-incorporated monomers. These polymers consist of phosphates with two R groups, one in the
backbone and the other as a side group. They have good biocompatibility and a
similarity to biomacromolecules such as DNA and RNA. Polyphosphoesters are
divided into two different classes: polyphosphonates with an alkyl or aryl R group
and polyphosphates with an alkoxy or aryloxy R group. Polyphosphoesters are
copolymerized with polyethers and polyesters to enhance their physical properties.
They have been studied as scaffolds for bone tissue regeneration [8].

References

1. E. Pişkin, Biodegradable polymers as biomaterials. J. Biomater. Sci. Polym. Ed. **6**(9), 775–795 (1995)
2. C.C. Chu, Biodegradable Polymeric Biomaterials: An Updated Overview, in *Biomedical Engineering Handbook*, 2nd edn., ed. by J.D. Bronzino (CRC Press, Boca Raton, Florida, 2000), pp. 1–22
3. B. Saad, P. Neuenschwander, G. Uhlschmid, U. Suter, New versatile, elastomeric, degradable polymeric materials for medicine. Int. J. Biol. Macromol. **25**(1–3), 293–301 (1999)
4. R. Yoda, Elastomers for biomedical applications. J. Biomater. Sci. Polym. Ed. **9**(6), 561–626 (1998)
5. R.M. Ginde, R.K. Gupta, In vitro chemical degradation of poly(glycolic acid) pellets and fibers. J. Appl. Polym. Sci. **33**(7), 2411–2429 (1987)
6. C.K.S. Pillai, C.P. Sharma, Review Paper: absorbable polymeric surgical sutures: chemistry, production, properties, biodegradability, and performance. J. Biomater. Appl. **25**(4), 291–366 (2010)
7. P.B. Maurus, C.C. Kaeding, Bioabsorbable implant material review. Operative Tech. Sports Med. **12**(3), 158–160 (2004)
8. B.D. Ulery, L.S. Nair, C.T. Laurencin, Biomedical applications of biodegradable polymers. J. Polym. Sci., Part B Polym. Phys. **49**(12), 832–864 (2011)
9. G.E. Luckachan, C.K.S. Pillai, Biodegradable polymers: a review on recent trends and emerging perspectives. J. Polym. Environ. **19**, 637–676 (2011)
10. H.R. Kricheldorf, Syntheses and application of polylactides. Chemosphere **43**(1), 49–54 (2001)
11. J.A. Cooper, H.H. Lu, F.K. Ko, J.W. Freeman, C.T. Laurencin, Fiber-based tissue-engineered scaffold for ligament replacement: design considerations and in vitro evaluation. Biomaterials **26**(13), 1523–1532 (2005)
12. H.H. Lu, J.A. Cooper Jr, S. Manuel, J.W. Freeman, M.A. Attawia, F.K. Ko, C.T. Laurencin, Anterior cruciate ligament regeneration using braided biodegradable scaffolds: in vitro optimization studies. Biomaterials **26**(23), 4805–4816 (2005)
13. M. Zilberman, K.D. Nelson, R.C. Eberhart, Mechanical properties and in vitro degradation of bioresorbable fibers and expandable fiber-based stents. J. Biomed. Mater. Res. Part B Appl. Biomater. **74**(2), 792–799 (2005)
14. J.C. Middleton, A.J. Tipton, Synthetic biodegradable polymers as orthopedic devices. Biomaterials **21**(23), 2335–2346 (2000)
15. J.E. Bergsma, F.R. Rozema, R.R.M. Bos, G. Boering, W.C. de Bruijn, A.J. Pennings, In vivo degradation and biocompatibility study of in vitro pre-degraded as-polymerized polylactide particles. Biomaterials **16**(4), 267–274 (1995)

16. C.C. Chu, Biodegradable polymeric biomaterials: an updated overview, in *Biomaterials: Principles and Applications*, ed. by J.B. Park, J.D. Bronzino (CRC Press, Florida, 2003), pp. 95–115
17. W. Amass, A. Amass, B. Tighe, A review of biodegradable polymers: uses, current developments in the synthesis and characterization of biodegradable polyesters, blends of biodegradable polymers and recent advances in biodegradation studies. Polym. Int. **47**(2), 89–144 (1998)
18. L.S. Nair, C.T. Laurencin, Biodegradable polymers as biomaterials. Prog. Polym. Sci. **32**(8–9), 762–798 (2007)
19. S. Li, S. McCarthy, Further investigations on the hydrolytic degradation of poly (DL-lactide). Biomaterials **20**(1), 35–44 (1999)
20. D.K. Gilding, A.M. Reed, Biodegradable polymers for use in surgery—polyglycolic/ poly(lactic acid) homo- and copolymers: 1. Polymer **20**(12), 1459–1464 (1979)
21. A.M. Reed, D.K. Gilding, Biodegradable polymers for use in surgery—poly(glycolic)/ poly(lactic acid) homo and copolymers: 2. In vitro degradation. Polymer **22**(4), 494–498 (1981)
22. X.L. Lu, Z.J. Sun, W. Cai, Z.Y. Gao, Study on the shape memory effects of poly(L-lactide-co-epsilon-caprolactone) biodegradable polymers. J. Mater. Sci. Mater. Med. **19**(1), 395–399 (2008)
23. W.L. Lee, P. Yu, M. Hong, E. Widjaja, S.C.J. Loo, Designing multilayered particulate systems for tunable drug release profiles. Acta Biomater. **8**, 2271–2278 (2012)
24. R.C. Mundargi, S. Srirangarajan, S.A. Agnihotri, S.A. Patil, S. Ravindra, S.B. Setty, T.M. Aminabhavi, Development and evaluation of novel biodegradable microspheres based on poly(d, l-lactide-co-glycolide) and poly(epsilon-caprolactone) for controlled delivery of doxycycline in the treatment of human periodontal pocket: in vitro and in vivo studies. J Control Release **119**(1), 59–68 (2007)
25. R.S. Bezwada, D.D. Jamiolkowski, I.Y. Lee, V. Agarwal, J. Persivale, S. Trenka-Benthin, M. Erneta, J. Suryadevara, A. Yang, S. Liu, Monocryl suture, a new ultra-pliable absorbable monofilament suture. Biomaterials **16**(15), 1141–1148 (1995)
26. J.-T. Hong, N.-S. Cho, H.-S. Yoon, T.-H. Kim, D.-H. Lee, W.-G. Kim, Preparation and characterization of biodegradable poly(trimethylenecarbonate-ε-caprolactone)-block-poly(p-dioxanone) copolymers. J. Polym. Sci. Part A Polym. Chem. **43**(13), 2790–2799 (2005)
27. K. Tomihata, M. Suzuki, N. Tomita, Handling characteristics of poly(L-lactide-co-epsilon-caprolactone) monofilament suture. Biomed. Mater. Eng. **15**(5), 381–391 (2005)
28. A. Steinbüchel, T. Lütke-Eversloh, Metabolic engineering and pathway construction for biotechnological production of relevant polyhydroxyalkanoates in microorganisms. Biochem. Eng. J. **16**(2), 81–96 (2003)
29. S.F. Williams, D.P. Martin, Applications of PHAs in medicine and pharmacy, in *Biopolymers for Medical and Pharmaceutical Applications*, vol. 1, ed. by A. Steinbüchel, R.H. Marchessault (Wiley-VCH, New Jersey, 2005), pp. 89–125
30. S.P. Valappil, S.K. Misra, A.R. Boccaccini, I. Roy, Biomedical applications of polyhydroxyalkanoates: an overview of animal testing and in vivo responses. Expert Rev. Med. Devices **3**(6), 853–868 (2006)
31. E.I. Shishatskaya, T.G. Volova, A.P. Puzyr, O.A. Mogilnaya, S.N. Efremov, Tissue response to the implantation of biodegradable polyhydroxyalkanoate sutures. J. Mater. Sci. Mater. Med. **15**(6), 719–728 (2004)
32. C. Ljungberg, G. Johansson-Ruden, K.J. Boström, L. Novikov, M. Wiberg, Neuronal survival using a resorbable synthetic conduit as an alternative to primary nerve repair. Microsurgery **19**(6), 259–264 (1999)
33. A. Hazari, G. Johansson-Rudén, K. Junemo-Bostrom, C. Ljungberg, G. Terenghi, C. Green, M. Wiberg, A new resorbable wrap-around implant as an alternative nerve repair technique. J Hand Surg Br **24**(3), 291–295 (1999)

34. L.N. Novikov, L.N. Novikova, A. Mosahebi, M. Wiberg, G. Terenghi, J.-O. Kellerth, A novel biodegradable implant for neuronal rescue and regeneration after spinal cord injury. Biomaterials 23(16), 3369–3376 (2002)
35. J.C. Knowles, F.A. Mahmud, G.W. Hastings, Piezoelectric characteristics of a polyhydroxybutyrate-based composite. Clinical Materials 8(1–2), 155–158 (1991)
36. C. Doyle, E.T. Tanner, W. Bonfield, In vitro and in vivo evaluation of polyhydroxybutyrate and of polyhydroxybutyrate reinforced with hydroxyapatite. Biomaterials 12(9), 841–847 (1991)
37. D.P. Martin, S.F. Williams, Medical applications of poly-4-hydroxybutyrate: a strong flexible absorbable biomaterial. Biochem. Eng. J. 16(2), 97–105 (2003)
38. K.-K. Yang, X.-L. Wang, Y.-Z. Wang, Poly(p-dioxanone) and its copolymers. J. Macromol. Sci. Polym. Rev. 42(3), 373 (2002)
39. H.L. Lin, C.C. Chu, D. Grubb, Hydrolytic degradation and morphologic study of poly-p-dioxanone. J. Biomed. Mater. Res. 27(2), 153–166 (1993)
40. S.W. Shalaby, Biomedical Polymers: Designed-To-Degrade Systems (Hanser Publishers, Munich, 1994)
41. R.S. Bezwada, S.W. Shalaby, H. Newman, A. Kafrawy, Bioabsorbable copolymers of p-dioxanone and lactide for surgical devices. Trans. Soc. Biomater. XIII, 194 (1990)
42. J. Kennedy, D. S. Kaplan, R. R. Muth, Absorbable composition. 522552006-Jul-(1993)
43. K. Gorona, S. Gogolewski, Novel biodegradable polyurethanes for medical applications, in Synthetic Bioabsorbable Polymers for Implants, ed. by C.M. Agrawal, J.E. Parr, S.T. Lin (ASTM special technical publication, UK, 2000), pp. 39–57
44. P. Gunatillake, R. Mayadunne, R. Adhikari, Recent developments in biodegradable synthetic polymers, in Biotechnology Annual Review, vol. 12, ed. by M. Raafat El-Gewely (Elsevier, Amsterdam, 2006), pp. 301–347
45. R.F. Storey, J.S. Wiggins, A.D. Puckett, Hydrolyzable poly(ester-urethane) networks from L-lysine diisocyanate and D, L-lactide/ε-caprolactone homo- and copolyester triols. J. Polym. Sci. Part A Polym. Chem. 32(12), 2345–2363 (2003)
46. J.Y. Zhang, E.J. Beckman, N.P. Piesco, S. Agarwal, A new peptide-based urethane polymer: synthesis, biodegradation, and potential to support cell growth in vitro. Biomaterials 21(12), 1247–1258 (2000)
47. M. Okada, Chemical syntheses of biodegradable polymers. Prog. Polym. Sci. 27(1), 87–133 (2002)
48. J. Heller, Polyorthoesters, in Handbook of Biodegradable Polymers, ed. by A.J. Domb, J. Kost, D.M. Wiseman (Harwood Academic Publishers, Australia, 1998), pp. 99–118
49. J. Heller, J. Barr, S. Ng, H.-R. Shen, K. Schwach-Abdellaoui, S. Emmahl, A. Rothen-Weinhold, R. Gurny, Poly(ortho esters)—their development and some recent applications. Eur. J. Pharm. Biopharm. 50(1), 121–128 (2000)
50. J. Heller, J. Barr, S.Y. Ng, K.S. Abdellauoi, R. Gurny, Poly(ortho esters): synthesis, characterization, properties and uses. Adv. Drug Deliv. Rev. 54(7), 1015–1039 (2002)
51. J. Heller, Development of poly(ortho esters): a historical overview. Biomaterials 11(9), 659–665 (1990)
52. J. Heller, S.Y. Ng, B.K. Fritzinger, K.V. Roskos, Controlled drug release from bioerodible hydrophobic ointments. Biomaterials 11(4), 235–237 (1990)
53. A. Göpferich, J. Tessmar, Polyanhydride degradation and erosion. Adv. Drug Deliv. Rev. 54(7), 911–931 (2002)
54. N. Kumar, R.S. Langer, A.J. Domb, Polyanhydrides: an overview. Adv. Drug Deliv. Rev. 54(7), 889–910 (2002)
55. H.B. Rosen, J. Chang, G.E. Wnek, R.J. Linhardt, R. Langer, Bioerodible polyanhydrides for controlled drug delivery. Biomaterials 4(2), 131–133 (1983)
56. K.E. Uhrich, A. Gupta, T.T. Thomas, C.T. Laurencin, R. Langer, Synthesis and characterization of degradable poly(anhydride-co-imides). Macromolecules 28(7), 2184–2193 (1995)

57. M.A. Attawia, K.E. Uhrich, E. Botchwey, M. Fan, R. Langer, C.T. Laurencin, Cytotoxicity testing of poly(anhydride-co-imides) for orthopedic applications. J. Biomed. Mater. Res. **29**(10), 1233–1240 (1995)

58. L. Erdmann, B. Macedo, K. Uhrich, Degradable poly(anhydride ester) implants: effects of localized salicylic acid release on bone. Biomaterials **21**(24), 2507–2512 (2000)

59. L. Erdmann, K.E. Uhrich, Synthesis and degradation characteristics of salicylic acid-derived poly(anhydride-esters). Biomaterials **21**(19), 1941–1946 (2000)

60. K. Whitaker-Brothers, K. Uhrich, Poly(anhydride-ester) fibers: role of copolymer composition on hydrolytic degradation and mechanical properties. J. Biomed. Mater. Res. Part A **70A**(2), 309–318 (2004)

61. S.I. Ertel, J. Kohn, Evaluation of a series of tyrosine-derived polycarbonates as degradable biomaterials. J. Biomed. Mater. Res. **28**(8), 919–930 (1994)

62. S.L. Bourke, J. Kohn, Polymers derived from the amino acid L-tyrosine: polycarbonates, polyarylates and copolymers with poly(ethylene glycol). Adv. Drug Deliv. Rev. **55**(4), 447–466 (2003)

63. K.A. Hooper, N.D. Macon, J. Kohn, Comparative histological evaluation of new tyrosine-derived polymers and poly (L-lactic acid) as a function of polymer degradation. J. Biomed. Mater. Res. **41**(3), 443–454 (1998)

64. V. Tangpasuthadol, S.M. Pendharkar, R.C. Peterson, J. Kohn, Hydrolytic degradation of tyrosine-derived polycarbonates, a class of new biomaterials. Part II: 3-yr study of polymeric devices. Biomaterials **21**(23), 2379–2387 (2000)

65. C. Meechaisue, R. Dubin, P. Supaphol, V.P. Hoven, J. Kohn, Electrospun mat of tyrosine-derived polycarbonate fibers for potential use as tissue scaffolding material. J. Biomater. Sci. Polym. Ed. **17**(9), 1039–1056 (2006)

66. S.L. Bourke, J. Kohn, M.G. Dunn, Preliminary development of a novel resorbable synthetic polymer fiber scaffold for anterior cruciate ligament reconstruction. Tissue Eng. **10**(1–2), 43–52 (2004)

67. S. Lakshmi, D. Katti, C. Laurencin, Biodegradable polyphosphazenes for drug delivery applications. Adv. Drug Deliv. Rev. **55**(4), 467–482 (2003)

68. A. Singh, N.R. Krogman, S. Sethuraman, L.S. Nair, J.L. Sturgeon, P.W. Brown, C.T. Laurencin, H.R. Allcock, Effect of side group chemistry on the properties of biodegradable L-alanine cosubstituted polyphosphazenes. Biomacromolecules **7**(3), 914–918 (2006)

69. M.T. Conconi, S. Lora, S. Baiguera, E. Boscolo, M. Folin, R. Scienza, P. Rebuffat, P.P. Parnigotto, G.G. Nussdorfer, In vitro culture of rat neuromicrovascular endothelial cells on polymeric scaffolds. J. Biomed. Mater. Res. Part A **71A**(4), 669–674 (2004)

70. L.S. Nair, S. Bhattacharyya, J.D. Bender, Y.E. Greish, P.W. Brown, H.R. Allcock, C.T. Laurencin, Fabrication and optimization of methylphenoxy substituted polyphosphazene nanofibers for biomedical applications. Biomacromolecules **5**(6), 2212–2220 (2004)

71. M.T. Conconi, S. Lora, A.M. Menti, P. Carampin, P.P. Parnigotto, In vitro evaluation of poly[bis(ethyl alanato)phosphazene] as a scaffold for bone tissue engineering. Tissue Eng. **12**(4), 811–819 (2006)

72. Q.-S. Zhang, Y.-H. Yan, S.-P. Li, T. Feng, Synthesis of a novel biodegradable and electroactive polyphosphazene for biomedical application. Biomed. Mater. **4**(3), 035008 (2009)

Chapter 5
Processing Parameters and the Rate of Resorption

Keywords Moisture · Temperature · Process-induced monomer · Sterilization techniques · Ethylene oxide · Radiation · Packaging

There are number of factors that affect the rate of resorption of hydrolytically sensitive biomaterials [1, 2]. These factors can be broadly classified into three groups under the headings: material properties, fabrication process parameters, and in vivo environmental conditions (Table 5.1).

Though all these factors are important for successful in vivo performance of biopolymers, the manufacturers who want to extrude fibers out of resorbable polymers would be more interested in factors such as spinning parameters, processing conditions, sterilization techniques, and packaging. The scope of the discussion in this chapter will thus be limited to these factors.

5.1 Effect of Processing Conditions and Spinning Parameters

Bioresorbable polymers can be processed using similar techniques to the ones that would normally be used for regular polymers, such as melting or dissolving in a solvent and extruding or spinning the filaments or continuous fibers, except that some precautions are required during the processing. This is because of the potential molecular weight reduction that is due to the hydrolytic sensitivity of the polymer. Such bioresorbable polymers usually have three degradation mechanisms during melt processing:

- Hydrolytic—this causes random chain scission.
- Mechanical—this is due to shear forces and also results in random chain scission.
- Thermal and Thermomechanical—these are two interrelated mechanisms resulting from high melt temperatures and/or long residence times leading to depolymerization.

C. R. Gajjar and M. W. King, *Resorbable Fiber-Forming Polymers for Biotextile Applications*, SpringerBriefs in Materials, DOI: 10.1007/978-3-319-08305-6_5, © The Author(s) 2014

Table 5.1 Classification of factors affecting rate of resorption of bioresorbable polymers

Material properties	Fabrication process	In vivo environment
Chemical composition of main chain and side groups	Type of process and process parameters	Site of implantation
Molecular weight and its distribution	Thermal treatment	pH of the fluids in the surrounding area
Monomer content	Annealing treatment	Enzymatic activities in the surrounding area
Crystallinity	Surface modification treatments	Infection and the growth of bacteria
Enantiomeric purity	External stresses	Proteins and lipids adhering to the implant surface
Residual moisture	Oxidation during processing and the resulting free radicals	Effect of synovial fluid in case of orthopedic implants
Additives and impurities	Process-induced monomers	Inflammatory reactions and the healing process
Hydrophilic–hydrophobic balance	Temperature and moisture of the surrounding environment	
Surface area, shape, and morphology of the polymer material	Sterilization technique	
	Packaging and storage conditions	

5.1.1 Effect of Moisture

Any residual moisture in the polymer as well as any atmospheric ambient moisture can have a detrimental effect on the rate of degradation of bioresorbable polymers during processing. Higher residual moisture can cause hydrolytic degradation during processing and alter the final properties of the fiber. Hence, care has to be taken to dry the polymer before processing and to prevent moisture from contacting the polymer during processing. It has been recommended to ensure that the residual moisture does not exceed 0.02 % to avoid excessive degradation during processing [3, 4]. Care should be taken even while drying above room temperature. Blanketing the hopper or material inlet with nitrogen or dried air would prevent moisture from entering the system. In addition, packaging the polymer in small quantities is recommended so that the material is used up quickly during processing once the package is opened to prevent ambient moisture absorption over time.

Care should be taken to achieve a balance and uniformity in the moisture content of the polymer. Bone dry resin requires higher processing temperatures, which are likewise harmful, while moist resin rapidly reduces the molecular weight during processing. So, the level of moisture content in the resin should be kept constant during the run to ensure uniform properties throughout.

Fig. 5.1 Molecular weight degradation for PLLA during injection molding as a function of residual moisture and the processing temperature [4]

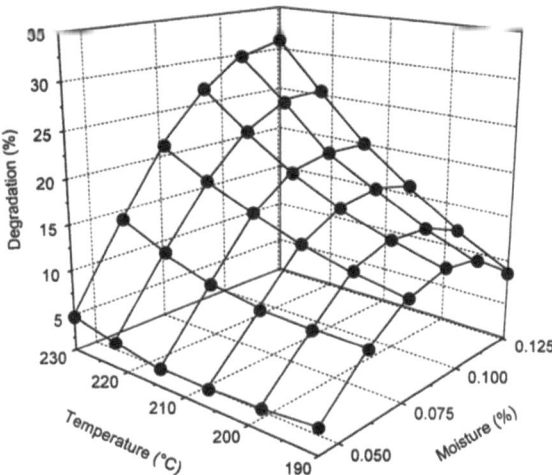

5.1.2 Effect of Temperature

High-processing temperatures increase the rate of degradation for bioresorbable polymers. At higher temperatures, there is a shift in equilibrium toward depolymerization, resulting in the formation of process-induced monomer. In turn, the presence of excess monomer has a plasticizing effect catalyzing the hydrolysis reaction. Even a small amount of residual moisture can have a catalytic effect on the hydrolysis of the polymer at high temperatures. The higher the residual moisture and the higher the processing temperature, the higher is the amount of degradation during processing (Fig. 5.1).

Another issue is the uniformity of the temperature distribution throughout the melt. In fact, the melt temperature is known to be highly non-uniform due to the low thermal conductivity of the viscous molten polymer. Large temperature differences (>50 °C) have been observed in different regions of the molten polymer within the same extruder, which results in non-uniform properties [5, 6].

5.1.3 Process-Induced Monomers

Process-induced monomer obviously increases the overall monomer content. Higher monomer content has a plasticizing effect, catalyzing the degradation process. Process-induced monomer affects the degradation kinetics, and results in the loss of mechanical properties. Important factors responsible for inducing monomer formation are high temperature, shear forces, residence time, and screw design and speed. The catalyst content of the polymer (usually Sn) also induces monomer formation [7]. However, it has been observed that for PLA the catalyst content does not play a significant role as long as the temperature is below 240 °C and the residence time is below 15 min. Resins with higher intrinsic viscosities (IV) require higher temperatures and thus result in more monomer formation [8]. In one study, Ella et al. demonstrated that the resorption time for PLA fibers was reduced due to an increase in the monomer content. It was found that after 9 weeks in vitro the fiber with 0.17 wt% monomer lost only 3 % of its strength, whereas the fiber with 1.24 wt% monomer lost as much as 65 % of its initial strength [9].

5.1.4 Extrusion Rate

The extrusion rate of a polymer is dependent on the rotational speed of the extruder which determines the residence time and shear rate of the polymer. The extrusion rate and the shear forces on the polymer have been found to affect its subsequent degradation profile [3, 4, 10]. One study found that a weaker, less drawn, and less crystalline fiber resorbed more quickly in PBS solution [10]. The extrusion speed determines the residence time at an elevated temperature, because it controls the time the material remains in the barrel of the extruder. Longer residence times (from 50 to 300 s) and higher shear rates result in higher rates of polymer degradation even at lower temperatures.

Thus in general, hydrolytically sensitive bioresorbable polymers need to be processed under the mildest conditions possible with rigorous exclusion of moisture from the polymer as well as the surrounding atmosphere. This is contradictory to the optimum melt-processing requirements. Many times, high temperatures are needed to reduce the melt viscosity, or higher pressures are needed to enable the polymer to be forced through the fine spinneret holes.

5.2 Effect of Sterilization Techniques, Packaging, and Storage Conditions

Packaging, sterilization, and storage conditions also play an important role in preventing the degradation of resorbable polymers. The susceptibility of resorbable polymers to hydrolysis presents problems when using the conventional methods of sterilization. Since these polymers are sensitive to heat, moisture, and radiation, choosing the right method of sterilization is an important consideration. Steam or dry-heat sterilization will degrade the device and hence cannot be used. Generally, bioresorbable implants are sterilized by γ-radiation, ethylene oxide (EtO), or other less-developed techniques such as low-temperature plasma sterilization or electron beam irradiation. Although effective, both γ-radiation and EtO have disadvantages. Radiation in the presence of air (oxygen) and moisture vapor, especially at doses higher than 2 Mrad, induces structural changes and degradation of the polymer chain, resulting in reduced molecular weight and affecting the mechanical properties and resorption time [11]. Hence, it is recommended that controlled environments, such as a nitrogen atmosphere and minimal residual moisture content, be combined with γ-radiation exposure.

Since the aliphatic polyesters such as PLA, PGA, and PDO are highly sensitive to radiation damage, these materials are usually sterilized by exposure to EtO. Unfortunately, the use of toxic and inflammable EtO presents a safety hazard of its own. EtO is a very reactive agent, and so, any measurable residuals of EtO are considered to put patients at risk. Thus, the implants sterilized by EtO must be degassed or aerated for an extended period of time (at least 24 h) to reduce EtO residuals to an acceptable level. Note that according to FDA requirements, the limits for residual EtO vary from 25 to 250 ppm, depending on the application of the implant. This can result in long-vacuum aeration times [12]. It is also important to note that in order for EtO to function as a bacteriocidal agent, it needs the presence of moisture vapor. So the necessary pre-humidification during EtO sterilization will also affect these hygroscopic polymers. Additionally, there is currently no method that can be used to sterilize tissue engineering scaffolds that are pre-seeded with viable cells or impregnated with growth factors. Such products are therefore manufactured under sterile conditions and used while still in a sterile condition [13].

Furthermore, special attention needs to be provided for packaging of resorbable implants. Since these polymers are hydrolytically unstable, the presence of moisture can degrade them in storage. Thus, the presence of any existing moisture must be eliminated, and ambient moisture vapor must be prevented from coming in contact with the polymer. However, since the polymers are naturally hygroscopic, eliminating the moisture and keeping the polymer free from moisture pose a major challenge. In practice, such polymers are quickly packaged after manufacture and generally double-bagged under an inert atmosphere or vacuum. Additionally, dry nitrogen can be purged into the packages to negate the effect of any residual moisture. The packaging material may involve airtight polymeric bags or sealed aluminum-backed plastic-foil pouches, but they must be resistant to water vapor permeability while at the same time being permeable to EtO vapor. The packed implants are typically stored in a freezer to minimize the effect of moisture. However, the package should be at room temperature before it is opened to minimize condensation, and it should be handled as little as possible at room temperature. Again, the shelf life stability is correlated with the degradation rate. For example, glycolides have a shorter shelf life than lactides. In one study, six different storage conditions were compared and it was shown that resorbable polymers remain stable even at room temperature for over 2 years as indicated by molecular weight loss when packaged in sealed and desiccated moisture-proof bags [14]. Final packaging consists of placing the packed material in an airtight moisture-proof container, which may be stored at lower temperatures below freezing. A desiccant may also be added to reduce the potential deleterious effect of moisture [15].

References

1. L. Fambri, C. Migliaresi, K. Kesenci, E. Piskin, Biodegradable Polymers, in *Integrated Biomaterials Science*, ed. by R. Barbucci (Kluwer Academic/Plenum, Berlin, 2002), pp. 119–187
2. C.C. Chu, Biodegradation Properties, in *Wound Closure Biomaterials and Devices*, ed by C.C. Chu, J.A.V. Fraunhofer, H.P. Greisler (CRC Press, United States, 1997), pp. 131–235
3. W. Michaeli, R. von Oepen, Processing of degradable polymers. ANTEC, 796–804 (1994)
4. R. von Oepen, W. Michaeli, Injection moulding of biodegradable implants. Clin. Mater. **10**(1–2), 21–28 (1992)
5. C. Rauwendaal, Finite element studies of flow and temperature evolution in single screw extruders. Plast. Rubber Compos. **33**(9–10), 390–396 (2004)
6. A.l. Kelly, E.C. Brown, P.D. Coates, The effect of screw geometry on melt temperature profile in single screw extrusion. Polym. Eng. Sci. **46**(12), 1706–1714 (2006)
7. T. Mori, H. Nishida, Y. Shirai, T. Endo, Effects of chain end structures on pyrolysis of poly(l-lactic acid) containing tin atoms. Polym. Degrad. Stab. **84**(2), 243–251 (2004)
8. K. Paakinaho, V. Ellä, S. Syrjälä, M. Kellomäki, Melt spinning of poly(l/d)lactide 96/4: effects of molecular weight and melt processing on hydrolytic degradation. Polym. Degrad. Stab. **94**(3), 438–442 (2009)
9. V. Ellä, L. Nikkola, M. Kellomäki, Process-induced monomer on a medical-grade polymer and its effect on short-term hydrolytic degradation. J. Appl. Polym. Sci. **119**(5), 2996–3003 (2011)
10. C. Golding, E. Ekevall, S. Wallace, R. Mather, The effect of degradation on the mechanical properties of biodegradable polylactide yarns and textiles, in *Medical Textiles and Biomaterials for Healthcare: Incorporating Proceedings of MEDTEX03 International Conference and Exhibition on Healthcare and Medical Textiles*, ed by S. Anand, J. Kennedy, M. Miraftab, S. Rajendran (Woodhead, Sawston, 2006), pp. 58–66
11. S.W. Shalaby, R.A. Johnson, Synthetic absorbable polyesters, in *Biomedical Polymers: Designed-To-Degrade Systems*, ed by S.W. Shalaby (Hanser Publishers, Munich, 1994), pp. 1–34
12. T. Zislis, S.A. Martin, E. Cerbas, J.R. Heath, J.L. Mansfield, J.O. Hollinger, A scanning electron microscopic study of in vitro toxicity of ethylene-oxide-sterilized bone repair materials. J. Oral Implantology **15**(1), 41 (1989)
13. J. Kohn, S. Abramson, R. Langer, Bioresorbable and Bioerodible Materials, in *Biomaterials Science—an Introduction to Materials in Medicine*, ed by B.D. Ratner, A.S. Hoffman, F.J. Schoen, J.E. Lemons, 2nd edn (Elsevier, Netherlands, 2004), pp. 115–127
14. C.T. Williams, J.C. Middleton, K.R. Sims, R.P. Swaim, D.R. Whitfield, J.C. Yarbrough, Long-term stability of biodegradable polymers, in *17th Southern Biomedical Engineering Conference*, 1998, p. 69
15. J.C. Middleton, A.J. Tipton, Synthetic biodegradable polymers as orthopedic devices. Biomaterials **21**(23), 2335–2346 (2000)

Chapter 6
Enzymatically Sensitive Fiber-Forming Bioresorbable Polymers

Keywords Enzymatic degradation · Biological hydrolysis · Biological oxidation · *Bombyx mori* · Polypeptides · Polysaccharides

In the previous chapters, we discussed different aspects of fiber-forming, hydrolytically initiated bioresorbable polymers for various biomedical applications. As mentioned earlier, there are essential requirements, in terms of biocompatibility and biofunctionality, which the polymer must meet in order to be used as a bioresorbable implant material. Due to the complex requirements, the number of commercially available and practically used hydrolytically initiated bioresorbable polymers has been restricted only to certain classes of polymers (mainly lactides and glycolides), and till now, only a few of them have been approved by the FDA for specific biomedical applications. Due to their hydrolytic sensitivity, these polymers present additional challenges during processing, manufacturing, sterilization, packaging, and storage.

One of the approaches to deal with this problem is to reduce the hydrolytic sensitivity of the biopolymers and induce in vivo resorption using alternative agents such as enzymes, temperature, pH, UV light, etc. Among various possible resorption-inducing agents, enzymes seem to be the most promising ones. Different enzymes are already present in the human body, which could be utilized to initiate and modulate the resorption of biopolymers in vivo, while avoiding the in vitro degradation of those polymers during manufacture. Hence, there is a lot of potential for enzymatically resorbable biopolymers. Many of the biopolymers from natural origin are enzymatically sensitive. These natural polymers have a structural similarity to components in host tissues and possess inherent advantages such as bioactivity, the ability to present receptor-binding ligands to cells for easy adhesion to the surface, susceptibility to cell-triggered proteolytic degradation, and natural remodeling [1]. Thus, the enzymatically resorbable natural biopolymers could be the ideal biomaterials. However, the strong inherent bioactivity of natural polymers has its own disadvantages such as a strong immunogenic response and the potential for rejection. In addition, there are other disadvantages like insufficient mechanical strength, difficulty in obtaining consistent and uniform quality, the need for purification, and the risk of disease transmission [2]. Furthermore, for

C. R. Gajjar and M. W. King, *Resorbable Fiber-Forming Polymers*
for Biotextile Applications, SpringerBriefs in Materials,
DOI: 10.1007/978-3-319-08305-6_6, © The Author(s) 2014

enzyme-catalyzed biopolymers, the rate of resorption would be dependent on enzyme concentration, which would vary with the site of implantation. Again, the characterization of a resorption profile is challenging since these polymers tend to give different results depending on whether they are exposed to in vivo or in vitro test conditions.

For these reasons, most of the research on enzymatic degradation of polymers has been focused on day-to-day plastics. Only some of the enzymatically resorbable natural and synthetic polymers have been investigated for biomedical applications. Lack of sufficient strength prevents the formation of fibers from such polymers, and hence, they have been limited for use in the form of hydrogels, coatings, nanofibers, and films for applications like drug delivery vehicles, biological coatings on implantable devices, and hemostatic wound dressings. In the next section, we will focus on the enzymatically resorbable biopolymers which have the potential of being spun into filaments and/or fibers for biomedical textile applications.

6.1 Requisites for Enzymatically Resorbable Polymers

The basic requirement for an enzymatically resorbable polymer is the presence of linkages, such as amides, which are susceptible to enzymatic attack. In the case of hydrolytically resorbable polymers, the rate of resorption is predominantly governed by the presence of hydrolysable bonds in the main chain structure, such as esters, acid anhydrides, and carbonates. However, enzymatically induced resorption depends on the extent of interaction between enzymes and the chemical bonds within the polymer. Thus, the polymer chains must be flexible enough to facilitate the interaction between the active site of the enzyme and the chemical bonds in the backbone chain. Higher chain flexibility can be achieved by those polymers which have lower crystallinity. Thus, less crystalline polymers are more susceptible to enzymatic resorption. Polymers with long repeating units are less likely to crystallize and therefore are likely to be more flexible and have a faster rate of enzymatic resorption [2]. Thus, a longer monomer chain is another prerequisite. This also explains why most enzymatically resorbable polymers are natural polymers rather than synthetic ones. Synthetic polymers tend to have shorter and regular repeat units, which enhances crystallization, makes the chains stiffer, and results in enzymes having less accessibility to the chemical bonds in the polymer chain. Another important consideration for enzymatic resorption of the polymer is the hydrophilic–hydrophobic balance. Since in vivo enzymatic resorption occurs in an aqueous media, the hydrophilic–hydrophobic character of the polymer and even segments within the polymer greatly affects its rate of bioresorption. One study has shown that those polymers containing both hydrophobic and hydrophilic segments have a faster rate of enzymatic resorption than those containing either entirely hydrophobic or entirely hydrophilic structures [3]. Table 6.1 summarizes the main requirements for enzymatically resorbable polymers.

Table 6.1 Requirements for enzymatically resorbable polymers

Preferable conditions for enzymatically resorbable polymers
Presence of chemical bonds which are susceptible to enzymatic attack
Accessibility between the chemical bonds of the polymer chain and the active site of the enzyme
Long, flexible monomer chains which reduce the level of crystallinity of the polymer
Hydrophilic as well as hydrophobic segments within the same polymer chain

6.2 Classification of Enzymatically Resorbable Biopolymers

Various enzymatically resorbable polymers are currently under in vitro and in vivo investigation with a view to applying them to clinical practice for different biomedical applications due to their unique advantages. However, their limited mechanical properties restrict their use. Most of these biopolymers are used in the form of hydrogels, nanofibers, or thin films. With some modifications, some of these biopolymers have the potential to be spun into fibers for biomedical textile applications. This section deals with the major enzymatically resorbable biopolymers. Table 6.2 shows the classification of enzymatically resorbable polymers that have been studied and/or are used for various biomedical applications [1, 2, 4].

6.2.1 Polypeptides

These are proteins and amino acid-based polymers having amide linkages. They are structurally similar to the components of tissue and hence find applications such as surgical sutures, hemostatic agents, tissue engineering scaffolds, and drug delivery devices. These polymers are hydrolytically stable but undergo rapid resorption in the presence of proteases. Various biopolymers in this group are described here.

6.2.1.1 Collagen

Collagen is the most abundant protein in the human body and is a major component of ligament, cartilage, tendon, skin, and bone. More than 28 different types of collagen have been identified to date, the most common being Types I–IV.

Collagen has been widely studied for biomedical applications such as bioresorbable sutures, artificial skin, wound dressings, and vascular grafts, due to its biocompatibility, mechanical properties, and degradability by enzymes like

Table 6.2 Classification of enzymatically resorbable biopolymers

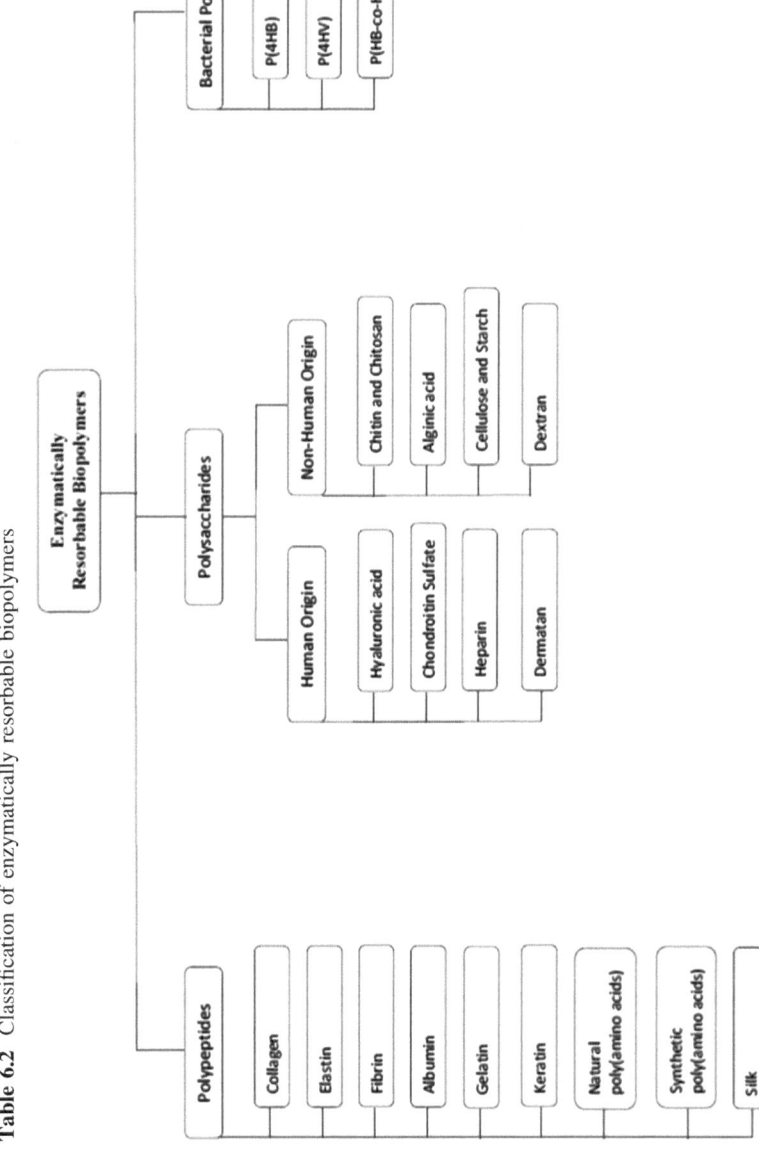

collagenases and metalloproteinases [5, 6]. Moreover, it is highly soluble in acidic solutions and hence easy to process in various forms such as sponges, tubes, sheets, and in powder form. Catgut is the commercially available enzymatically resorbable collagen suture. Collagen plays a major role in the initiation of the coagulation cascade and is thrombogenic. Hence, it is useful as a hemostatic agent in the form of either microfibers or a hydrogel for wound dressings. Collagen shows good cell attachment, proliferation, and differentiation and hence a potential biomaterial for tissue engineering scaffolds [7]. Absorbable collagen sponges have been investigated as scaffold materials for accelerated tissue regeneration [8, 9].

To improve collagen's potential as a biomaterial, it has been modified or combined with other resorbable polymers. Modifications like cross-linking, addition of bioactive molecules, and enzymatic pretreatment have resulted in novel collagen-based materials with improved functionality [10, 11]. Moreover, to facilitate the formation of fibers for biomedical textile applications, composite materials combining collagen with other resorbable polymers like PLA, PLGA, and PCL have been studied extensively [12–15].

Currently, collagen for medical use is obtained exclusively from precisely identified, controlled, and monitored bovine and porcine sources. This has been mandated by the major national regulatory agencies around the world that are required to protect patients from the risk of infectious disease transmission, particularly from Creutzfeldt–Jakob disease (bovine spongiform encephalopathy) otherwise known as 'mad cow disease.' Other limiting factors for the widespread clinical use of pure collagen include its mild immunogenicity and high cost [4].

6.2.1.2 Elastin and Elastin-Like Peptides

Elastin is a major protein component of vascular and lung tissue and is responsible for imparting highly elastic properties. Elastin is a highly cross-linked insoluble polymer. It has been found to have minimum interaction with platelets, which makes it an attractive biological material for coating synthetic vascular grafts [16]. However, elastin elicits an immune response similar to that for collagen implants, which limits its applicability to certain biomedical end uses.

To overcome these inherent limitations of elastin, synthetic elastins and elastin-like polypeptides (ELP) have been developed. They have good biocompatibility and non-immunogenic properties, and the degradation by-products are natural amino acids. ELPs show an interesting property of an inverse temperature transition (ITT), whereby the molecular organization changes from a disordered state to an ordered state at temperatures above 25 °C. This makes them promising materials for smart drug delivery systems [17]. ELPs have a shear modulus similar to normal cartilage (about 1.7 kPa), making them an ideal biomaterial candidate for tissue engineering of cartilage and other soft tissues [18].

6.2.1.3 Fibrin/Fibrinogen

Fibrin is a biopolymer similar to collagen. It is an important component of the natural blood clotting process. Fibrin has long been used as a biopolymer due to its excellent biocompatibility and rapid bioresorption. It is widely used as a sealant to induce thrombosis and prevent blood loss from wounds. Fibrin also facilitates cell adhesion and proliferation.

A novel biomaterial has been developed by reacting fibrin monomer with elastin. The resulting elastin–fibrin (EF) material acts as a connective tissue matrix. EF materials have been used in surgery to reinforce and repair damaged connective tissues in the human intestinal tract [19–21].

6.2.1.4 Albumin

Albumin is a water-soluble protein and a major component of human blood plasma. It can be easily processed into various forms like membranes, microspheres, nanofibers, and nanospheres [22]. Studies have shown that almost all tissues in the human body have the ability to degrade albumin [23]. For this reason, albumin is usually cross-linked so as to slow its rate of resorption or lysis. Albumin has been studied extensively for applications like surgical adhesives, coating materials for cardiovascular devices, and drug delivery vehicles [24, 25].

6.2.1.5 Natural Poly(amino acids)

Natural poly(amino acids) are ionic polymers having only one type of amino acid joined together with amide linkages. When used as a biomaterial, poly(γ-glutamic acid) (γPGA) and poly(L-lysine) are the two most commonly studied natural poly(amino acids).

γPGA is a water-soluble homopolyamide produced by microbial fermentation. The high functionality of γPGA makes it an attractive biomaterial for drug delivery and for developing bioactive scaffolds for tissue engineering. An interesting property of γPGA is its thermosensitivity. One particular study has shown that γPGA exhibits unique changes in its level of hydration properties with a change in temperature [26]. This makes them potentially attractive for nanoengineered intelligent biomaterials. However, γPGA has inferior mechanical properties. It can be blended with other biomaterials, like PLA, PLGA, and PCL, to form strong and hydrophilic composites. Commercial interest in using γPGA is not widespread due to its limited availability.

Poly(L-lysine) is also synthesized by bacteria and is being evaluated for its use as a scaffold material for tissue engineering and as a drug delivery vehicle. It has antibacterial, antiviral, and antitumor properties. However, the polymer has a highly positive charge which causes it to be cytotoxic and limits its wider application.

6.2.1.6 Synthetic Poly(amino acids)

Due to their similarity to naturally occurring proteins, several poly(amino acids) have been synthesized and studied. However, their high crystallinity, slow rate of degradation, inferior mechanical properties, and immunogenic response prevent their widespread clinical use. Only two synthetic poly(amino acids) have been found to be promising biomaterials.

Poly(L-glutamic acid) (L-PGA) is synthesized from naturally occurring L-glutamic acid residues. It differs from γPGA in that its amide linkage is made with the α-carbon amine group instead of γ-carbon amine group. This makes L-PGA more flexible than γPGA. L-PGA is very biocompatible and non-immunogenic and susceptible to degradation by lysosomal enzymes. It has shown promising results for use in drug delivery systems and as a resorbable contrast agent [27]. To improve its properties and potential for biomedical applications, L-PGA has been copolymerized with PLA, PCL, and PEG. Some attempts have been made to develop synthetic copolymeric polypeptide fibers, and a partly esterified L-PGA has been developed and evaluated as a surgical suture [28].

Poly(aspartic acid) (PAA) is another promising biomaterial synthesized from aspartic acid. It is a highly water-soluble ionic polymer which is readily degraded by lysosomal enzymes. It can be readily prepared as a hydrogel for various biomedical applications, and it has also been copolymerized with PLA, PCL, and PEG to improve its mechanical performance and reduce its hydrophilicity and rate of degradation.

6.2.1.7 Silk

Silk is an interesting biomaterial that has been used since ancient times. It is a protein polymer (consisting of various amino acids) that is spun into fibers by insects like the silkworm, spider, scorpion, mites, and flies. Depending on the source and the arrangement of amino acids, there are a variety of silks, each having specific properties. Some evolutionary advanced species of insects are capable of spinning as many as nine or more varieties of silk for different purposes such as cocoon construction, lines for prey capture, safety lines or draglines, web construction, and adhesion [29]. The most widely characterized silks include those from the domesticated silkworm (*Bombyx mori*) and from spiders (*Nephila clavipes* and *Araneus diadematus*).

Conventionally, cultivated silk from the silkworm (*B. mori*) has been the dominant variety for commercial applications since sericulture has been part of the Chinese textile industry for 5,000 years. Spider silks have not been commercialized primarily due to the predatory nature of spiders and the relatively low levels of production. However, silks have superior mechanical properties (Table 6.3) [30].

While silk from the domesticated silkworm (*B. mori*) has been used as a biomaterial, primarily as a suture, the biological response to sericin has raised

Table 6.3 Mechanical properties of silks [30]

Material origin	Tensile strength (MPa)	Modulus (GPa)	Elongation at break (%)
Bombyx mori silk (with sericin)	500	5–12	19
Bombyx mori silk (without sericin)	610–690	15–17	4–16
Spider silk	875–972	11–13	17–18

questions about its biocompatibility. Sericin is the glue-like protein that holds the two spun filaments (or brins) of silk together at the time the silkworm larva (or caterpillar) creates its cocoon. In one study, silkworm silk still containing sericin evoked a fibroblastic response when used as a suture material for attaching nerves in a rabbit model [31]. It is therefore essential to remove all the accompanying sericin component by an alkali degumming or scouring process prior to using any cultivated silk as a surgical suture or other implantable device [32].

In contrast, spider silk is devoid of sericin and hence does not evoke the same biological or immunological reactions. Thus, spider silk has better biocompatibility and is a preferred biomaterial for suture applications. It has also been studied as a material for regenerative nerve conduits to promote peripheral nerve regeneration [33]. Silk's unique mechanical properties coupled with its ability to be fabricated into different textile structures enable its use in tissue engineering scaffolds that mimic the mechanical properties of native tissues. For example, silk filaments have been converted into a braided rope structure that acts as a scaffold for the regeneration of anterior cruciate ligaments (ACL) [34].

There is a lot of ambiguity as to whether silk is resorbable or not. US Pharmacopeia classifies silk as non-degradable material. This is because the US Pharmacopeia defines absorbable material as one that loses most of its tensile strength in vivo within 60 days post-implantation. However, the literature shows that silk undergoes proteolytic degradation over a longer period of time, which is 1–2 years. Studies have indicated that silk fibers lose most of their tensile strength within the first year in vivo and resorb completely within 2 years of implantation [30, 35].

6.2.2 Polysaccharides

Polysaccharides are macromolecules having glycosidic linkages. Their bioresorption and bioactivity make them attractive biomaterials. They show biological functions ranging from cell signaling to immune recognition. Moreover, they are abundantly available in nature. Because of these attributes, they are the most extensively studied natural biomaterials. Polysaccharides used in biomedical applications can be classified as those derived from human origin and those from non-human sources.

6.2.2.1 Polysaccharides from Human Origin

Hyaluronic Acid

Hyaluronic acid (HA) is a naturally occurring glycosaminoglycan consisting of alternating units of N-acetyl-D-glucosamine and glucuronic acid. It is water soluble and forms highly viscous solutions. HA is found as a major component of synovial fluid and vitreous humor. Traditionally, HA has been derived from animal sources, but recent advances in microbiological techniques have led to the production of the first animal-free sodium hyaluronate which is synthesized by the bacteria, *Bacillius subtilis*.

At body temperature, the HA homopolymer is a fluid and therefore is not a structural biomaterial. To overcome this limitation, HA has been cross-linked with ethyl esters or benzyl esters to form hydrogels, which can undergo resorption during periods ranging from 1–2 weeks to 4–5 months [36–38]. This biomaterial is commercially available as HYAFF® (Fidia Farmaceutici). HYAFF® hydrogels have been shown to be quite versatile and can be fabricated into sheets, membranes, sponges, tubes, fibers, gauzes, or nonwoven fabrics [39–41]. The hydrogel has been evaluated extensively for use as an adhesion barrier, wound dressing, and scaffold for the regeneration of the trachea, articular cartilage, blood vessels, and nerve tissue. It has also shown promising results as a drug delivery vehicle. Composites of HA with PLA, PLGA, and PCL have been developed to produce biomaterials with enhanced mechanical properties.

6.2.2.2 Chondroitin Sulfate

Chondroitin sulfate (CS) is another glycosaminoglycan which is a promising biomaterial. It is a major component of the natural, hydrophilic wound-healing matrix produced by fibroblasts during the normal inflammatory and foreign body response to injury. CS has been shown to stimulate the metabolic response of cartilage and possesses anti-inflammatory properties [42, 43]. It also facilitates intracellular signaling, cell recognition, and the connection of cells to the extracellular matrix components at the wound site [44]. Due to its biocompatibility, non-immunogenicity, and pliability, CS has been investigated for wound-dressing applications [45]. Studies have shown that successful cartilage regeneration can be achieved through the use of a tissue-engineered implant that causes the correct phenotypic development of the seeded chondrocytes. Since CS plays an important role in regulating the expression of the chondrocyte phenotype, it has been extensively studied as a scaffold for cartilage tissue engineering. Composites of CS and other polymers like PLA, PEG, and PCL have also been evaluated with the objective of delineating the optimum resorbable bioactive scaffolds for cartilage tissue regeneration.

6.2.2.3 Polysaccharides from Non-human Origin

Chitin and Chitosan

Chitin is a naturally abundant polysaccharide present in the exoskeleton of many arthropods, such as shrimp and crab. It is also found as the structural pen of the squid and in certain types of mushroom. Chitin is structurally similar to HA and also has been shown to have good wound-healing properties [46]. Chitin can be degraded by chitinase. It has low immunogenicity, and its mechanical strength is high enough to permit the fabrication of functional fibers. Chitin fibers have been investigated for applications such as wound dressings, artificial skin, and absorbable sutures [47–49]. However, chitin is insoluble in many common solvents, which limits its ease of processing and commercial potential.

To overcome this issue, chitin is converted to chitosan by deacetylation, which is then water soluble. Chitosan is degraded by a number of enzymes including chitosanase, lysozyme, and papain. Chitosan shows biostimulating activities in the healing process, inhibits fibroplasia in wound healing, and promotes tissue growth [50, 51]. In addition, chitosan is water absorptive, oxygen permeable, and hemostatic. Chitosan's intrinsic antibacterial property, minimal foreign body reaction, and biocompatible degradation by-products also help in rapid wound healing. In fact, it has been evaluated as a wound-dressing material for more than 20 years [52, 53]. For example, HemCon® (HemCon Medical Technologies) is an FDA-approved, commercial wound dressing based on chitosan. To improve its mechanical properties, chitosan has been either cross-linked or combined with other biopolymers like PLA, PLGA, PEG, and collagen and fabricated into films, membranes, sponges, nanoparticles, fibers, and gels [54–56].

Due to its hydrophilicity and cationic nature, chitosan has been investigated as a drug delivery vehicle for the targeted delivery of chemotherapeutics, antibiotics, as well as DNA and proteins. The ability to form porous matrices makes chitosan an attractive material for use as a tissue engineering scaffold. Chitosan and its family of composites with other biopolymers has been fabricated into fibrous materials for use as tissue engineering scaffolds for the regeneration of bone, ligaments, tendons, nerve, and skin [57].

Alginic Acid

Alginic acid is a linear block copolymer synthesized in the cell wall and intercellular spaces of brown algae. It gives flexibility and strength to marine plants. The high functionality of alginic acid makes it an attractive material for biomedical applications. It has been studied for applications such as a drug delivery device, wound dressing, and tissue engineering scaffold. Since alginate, by itself, is mechanically weak, it has been copolymerized with PLGA, collagen, and chitosan to improve its mechanical performance. These composite materials have been fabricated into films, sponges, fibers, and gels and used as scaffolds for regenerating bone, cartilage, liver, nerve, and vascular tissues [58].

Despite having some superior properties for biomedical applications, alginate has two main limitations: its rate of in vivo resorption is slow, and it has poor cellular adhesion. Hence, alginate is often irradiated with gamma radiation or oxidized by periodate to increase its rate of resorption, and chemical modification of the alginate side groups with bioactive molecules is currently being investigated to improve cellular adhesion.

Cellulose and Starch

Cellulose is a major naturally occurring storage and structural polymer in plants. Cellulose and its derivatives have been widely investigated for various biomedical applications. Although pure cellulose is not readily resorbable in the body, it can be chemically modified to cause chain scission of the structure and hence be made bioresorbable. Oxidized cellulose is one such example. Surgicel® is a commercial oxidized cellulose, which is available in the form of gauze pads for use as resorbable hemostatic wound dressings.

Starch is another polysaccharide that is being studied for biomedical applications in the form of gels, films, and microspheres. Its poor mechanical strength prevents its use in the form of fibers.

6.2.3 Bacterial Polyesters

Polyhydroxyalkanoates (PHAs) such as poly(4-hydroxybutyric acid) (P(4HB)), poly(4-hydroxyvalerate) (P(4HV)), and their copolymers are bacterial polyesters synthesized by microbial fermentation. They have been described in detail in the earlier section on hydrolytically resorbable biopolymers. They are also found to be resorbable by enzymatic action in vivo. This particular family of bacterial polyesters is one of the most promising biomaterials currently under investigation.

6.3 Mechanism of Enzymatic Resorption

Each of the enzymatically resorbable biopolymers has certain specific enzymes that can attack the chemical bonds of the polymeric material, leading to its degradation and resorption. Different enzymes have different actions. Some enzymes change the substrate through a free radical mechanism, while others follow an alternative chemical route. Typical examples are biological oxidation and biological hydrolysis.

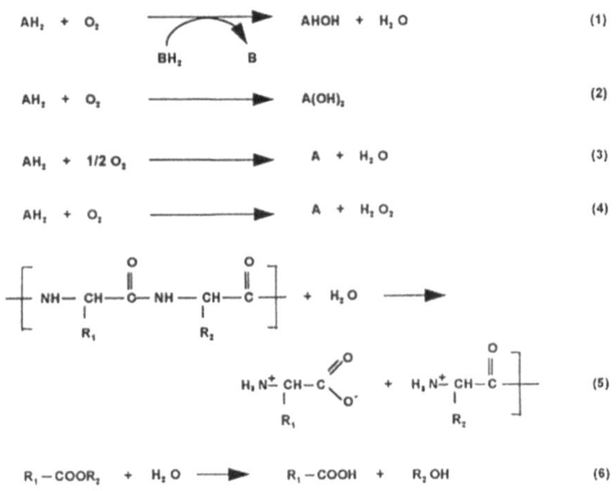

Fig. 6.1 Biological oxidation and hydrolysis by enzymes [62]

$$AH_2 \; + \; O_2 \xrightarrow{\quad BH_2 \qquad B \quad} AHOH \; + \; H_2O \qquad (1)$$

$$AH_2 \; + \; O_2 \longrightarrow A(OH)_2 \qquad (2)$$

$$AH_2 \; + \; 1/2\,O_2 \longrightarrow A \; + \; H_2O \qquad (3)$$

$$AH_2 \; + \; O_2 \longrightarrow A \; + \; H_2O_2 \qquad (4)$$

$$R_1\!-\!COOR_2 \; + \; H_2O \longrightarrow R_1\!-\!COOH \; + \; R_2OH \qquad (6)$$

6.3.1 Biological Oxidation

Polymeric biomaterials may be degraded by chemical and/or enzymatic oxidation when exposed to body fluids and tissues. During the inflammatory response to foreign materials, inflammatory cells, especially leukocytes and macrophages, produce highly reactive oxygen species such as superoxides (O_2^-), hydrogen peroxide (H_2O_2), nitric oxide (NO), and hypochlorous acid (HOCl). The oxidative effect of these species may cause polymer chain scission and initiate the process of degradation [59–61].

Several enzymes, like cytochromoxidase, can react directly with oxygen in the respiratory system, while in other cases, oxygen is incorporated into the substrate. The enzyme hydroxylase, also called monooxynase, catalyzes the insertion of a single atom of oxygen into the substrate in the form of a hydroxyl group (Fig. 6.1, Eq. 1). Oxygenases, also called dioxygenases, catalyze the insertion of a whole oxygen molecule into the substrate as a dihydroxy derivative or carbonyl (–CO) or carboxyl (–COO) group (Fig. 6.1, Eq. 2). Another type of biological oxidation is when the oxygen molecule is not actually incorporated into the substrate, but rather, it functions as an electron acceptor. Oxidases are such enzymes which either produce H_2O (Fig. 6.1, Eq. 3) or H_2O_2 (Fig. 6.1, Eq. 4) [62].

6.4 Biological Hydrolysis

Enzymes can also catalyze hydrolysis reactions which lead to the resorption of polymeric biomaterials. For example, proteolytic enzymes (proteases) catalyze the hydrolysis of peptide bonds (Fig. 6.1, Eq. 5) as well as the hydrolysis of ester bonds (Fig. 6.1, Eq. 6) [62].

6.4.1 Factors Affecting Enzymatic Resorption

The enzymatic resorption of a biomaterial involves the interaction between enzymes and the chemical bonds in the polymer chain. Mechanistically, this process can be divided into four sequential steps [61]:

1. Diffusion of the enzyme from the bulk solution to the solid surface
2. Adsorption of the enzyme onto the substrate
3. Catalysis of a chain scission reaction
4. Diffusion of the soluble degradation by-products from the substrate into the surrounding environment.

The rates of adsorption and chain scission are affected by physicochemical properties of the substrate, such as the molecular weight, chemical composition, crystallinity, and surface area, and also by the inherent characteristics of the enzyme which can be measured in terms of its activity, stability, concentration, amino acid composition, and conformation. Moreover, environmental conditions such as pH and temperature also influence the activity of enzymes. The presence of stabilizers, activators, or inhibitors released from the polymer during the degradation process or additives that are leached out may also affect enzyme activity. Chemical modification of biopolymers may also affect the rate of enzymatic resorption since, depending on the degree of chemical modification, it may prevent the enzyme from recognizing the polymeric substrate. The rate of enzymatic resorption is limited by an enzyme saturation point. Beyond this enzyme concentration, no further increase in the rate of resorption is observed even when more enzyme is added.

On the whole, the complex and variable chemical nature of body fluids and tissues around an implant, the variability of polymer materials themselves, and the inherent variability of biological systems make biomaterial resorption by an enzymatic mechanism a complex process.

6.5 Enzymes for Synthetic Biopolymers

Most of the enzymatically resorbable biopolymers discussed above have a natural origin. Nevertheless, in order to improve the resorption profile of the synthetic hydrolytically resorbable biopolymers, studies have been undertaken to identify those enzymes that are able to catalyze the resorption of synthetic biopolymers.

Synthetic biopolymers like PLLA and PGA are synthesized from natural materials like corn, and hence, they are expected to be degraded by enzymes. The enzymatic hydrolysis of PLLA has been studied using proteinase K, and it has been shown that this enzyme is able to significantly accelerate the hydrolysis of PLLA [63, 64]. In addition, certain lipase enzymes have been found to enhance the bioresorption of PCL [65]. Thus, there is the possibility for enhancing the bioresorption of synthetic biopolymers using specific enzymes.

References

1. L.S. Nair, C.T. Laurencin, Biodegradable polymers as biomaterials. Prog. Polym. Sci. **32**(8–9), 762–798 (2007)
2. T. Hayashi, Biodegradable polymers for biomedical uses. Prog. Polym. Sci. **19**(4), 663–702 (1994)
3. W.J. Bailey, Y. Okamoto, W.C. Kuo, T. Narita, in *Proceedings of 3rd International Biodegradation Symposium*, ed. by J.M. Sharpley, A.M. Kaplan (Applied Science Publishers, London, 1976), p. 765
4. B.D. Ulery, L.S. Nair, C.T. Laurencin, Biomedical applications of biodegradable polymers. J. Polym. Sci., Part B: Polym. Phys. **49**(12), 832–864 (2011)
5. T. Okada, T. Hayashi, Y. Ikada, Degradation of collagen suture in vitro and in vivo. Biomaterials **13**(7), 448–454 (1992)
6. I.V. Yannas, J.F. Burke, Design of an artificial skin. I. Basic design principles. J. Biomed. Mater. Res. **14**(1), 65–81 (1980)
7. R.J. Bellucci, D. Wolff, Experimental stapedectomy with collagen sponge implant. The Laryngoscope **74**(5), 668–688 (1964)
8. P.K. Narotam, S. José, N. Nathoo, C. Taylon, Y. Vora, Collagen matrix (DuraGen) in dural repair: analysis of a new modified technique. Spine **29**(24), 2861–2867; discussion 2868–2869 (2004)
9. X. Duan, C. McLaughlin, M. Griffith, H. Sheardown, Biofunctionalization of collagen for improved biological response: scaffolds for corneal tissue engineering. Biomaterials **28**(1), 78–88 (2007)
10. Y.M. Bastiaansen-Jenniskens, W. Koevoet, A.C.W. de Bart, J.C. van der Linden, A.M. Zuurmond, H. Weinans, J.A.N. Verhaar, G.J.V.M. van Osch, J. Degroot, Contribution of collagen network features to functional properties of engineered cartilage. Osteoarthr. Cartil. **16**(3), 359–366 (2008)
11. S. Torres-Giner, J.V. Gimeno-Alcañiz, M.J. Ocio, J.M. Lagaron, Comparative performance of electrospun collagen nanofibers cross-linked by means of different methods. ACS Appl. Mater. Interfaces **1**(1), 218–223 (2009)
12. J.G. Cho-Hong, S. Sahoo, Development of hybrid polymer scaffolds for potential applications in ligament and tendon tissue engineering. Biomed. Mater. (Bristol, England) **2**(3), 169–173 (2007)
13. Y. Tanaka, H. Yamaoka, S. Nishizawa, S. Nagata, T. Ogasawara, Y. Asawa, Y. Fujihara, T. Takato, K. Hoshi, The optimization of porous polymeric scaffolds for chondrocyte/atelocollagen based tissue-engineered cartilage. Biomaterials **31**(16), 4506–4516 (2010)
14. C. Xu, W. Lu, S. Bian, J. Liang, Y. Fan, X. Zhang, Porous collagen scaffold reinforced with surfaced activated PLLA nanoparticles. Sci. World J. **2012** (2012)
15. Y. Tatekawa, N. Kawazoe, G. Chen, Y. Shirasaki, H. Komuro, M. Kaneko, Tracheal defect repair using a PLGA-collagen hybrid scaffold reinforced by a copolymer stent with bFGF-impregnated gelatin hydrogel. Pediatr. Surg. Int. **26**(6), 575–580 (2010)
16. K.A. Woodhouse, P. Klement, V. Chen, M.B. Gorbet, F.W. Keeley, R. Stahl, J.D. Fromstein, C.M. Bellingham, Investigation of recombinant human elastin polypeptides as non-thrombogenic coatings. Biomaterials **25**(19), 4543–4553 (2004)
17. S.M. Mithieux, J.E.J. Rasko, A.S. Weiss, Synthetic elastin hydrogels derived from massive elastic assemblies of self-organized human protein monomers. Biomaterials **25**(20), 4921–4927 (2004)
18. H. Betre, S.R. Ong, F. Guilak, A. Chilkoti, B. Fermor, L.A. Setton, Chondrocytic differentiation of human adipose-derived adult stem cells in elastin-like polypeptide. Biomaterials **27**(1), 91–99 (2006)
19. F. Lefebvre, F. Drouillet, A.M.S. de Larclause, M. Aprahamian, D. Midy, L. Bordenave, M. Rabaud, Repair of experimental arteriotomy in rabbit aorta using a new resorbable elastin—fibrin biomaterial. J. Biomed. Mater. Res. **23**(12), 1423–1432 (1989)

20. C. Barbié, C. Angibaud, T. Darnls, F. Lefebvre, M. Rabaud, M. Aprahamian, Some factors affecting properties of elastin-fibrin biomaterial. Biomaterials **10**(7), 445–448 (1989)

21. D. Collet, F. Lefebvre, C. Quentin, M. Rabaud, In vitro studies of elastin-fibrin biomaterial degradation: preservative effects of protease inhibitors and antibiotics. Biomaterials **12**(8), 763–766 (1991)

22. Y. Dror, T. Ziv, V. Makarov, H. Wolf, A. Admon, E. Zussman, Nanofibers made of globular proteins. Biomacromolecules **9**(10), 2749–2754 (2008)

23. B.H. Prinsen, M.G.M. de Sain-van der Velden, Albumin turnover: experimental approach and its application in health and renal diseases. Clinica. Chimica. Acta **347**(1–2), 1–14 (2004)

24. V. Tuan Giam Chuang, U. Kragh-Hansen, M. Otagiri, Pharmaceutical strategies utilizing recombinant human serum albumin. Pharm. Res. **19**(5), 569–577 (2002)

25. M. Uchida, A. Ito, K.S. Furukawa, K. Nakamura, Y. Onimura, A. Oyane, T. Ushida, T. Yamane, T. Tamaki, T. Tateishi, Reduced platelet adhesion to titanium metal coated with apatite, albumin-apatite composite or laminin-apatite composite. Biomaterials **26**(34), 6924–6931 (2005)

26. T. Shimokuri, T. Kaneko, M. Akashi, Specific thermosensitive volume change of biopolymer gels derived from propylated poly(γ-glutamate)s. J. Polym. Sci., Part A: Polym. Chem. **42**(18), 4492–4501 (2004)

27. G. Zhang, R. Zhang, X. Wen, L. Li, C. Li, Micelles based on biodegradable poly(L-glutamic acid)-b-Polylactide with paramagnetic Gd Ions chelated to the shell layer as a potential nanoscale MRI-visible delivery system. Biomacromolecules **9**(1), 36–42 (2008)

28. T. Miyamae, S. Mori, Y. Takeda, Filaments and surgical sutures of poly(L-glutamic acid) partly esterified with lower alkanols. 337106927-Feb-1968 (1968)

29. D.L. Kaplan, S. Fossey, C. Viney, W. Muller, Self-organization (assembly) in biosynthesis of silk fibers: a hierarchical problem, in *Hierarchically Structured Materials*, vol. 255, ed by I. A. Aksay, E. Baer, M. Sarikaya, D. Tirrell, Materials Res Symposium Proceedings, pp. 19–29 (1992)

30. G.H. Altman, F. Diaz, C. Jakuba, T. Calabro, R.L. Horan, J. Chen, H. Lu, J. Richmond, D.L. Kaplan, Silk-based biomaterials. Biomaterials **24**(3), 401–416 (2003)

31. J.C. DeLee, M.T. Smith, D.P. Green, The reaction of nerve tissue to various suture materials: a study in rabbits. J. Hand. Surg. Am. **2**(1), 38–43 (1977)

32. X. Yang, L. Wang, G. Guan, M.W. King, Y. Li, L. Peng, Y. Guan, X. Hu, Preparation and evaluation of bicomponent and homogeneous polyester silk small diameter arterial prostheses. J. Biomater. Appl. **28**(5), 676–687 (2014)

33. C. Allmeling, A. Jokuszies, K. Reimers, S. Kall, C.Y. Choi, G. Brandes, C. Kasper, T. Scheper, M. Guggenheim, P.M. Vogt, Spider silk fibres in artificial nerve constructs promote peripheral nerve regeneration. Cell Prolif. **41**(3), 408–420 (2008)

34. G.H. Altman, R.L. Horan, H.H. Lu, J. Moreau, I. Martin, J.C. Richmond, D.L. Kaplan, Silk matrix for tissue engineered anterior cruciate ligaments. Biomaterials **23**(20), 4131–4141 (2002)

35. Y. Cao, B. Wang, Biodegradation of silk biomaterials. Int. J. Mol. Sci. **10**(4), 1514–1524 (2009)

36. L. Benedetti, R. Cortivo, T. Berti, A. Berti, F. Pea, M. Mazzo, M. Moras, G. Abatangelo, Biocompatibility and biodegradation of different hyaluronan derivatives (Hyaff) implanted in rats. Biomaterials **14**(15), 1154–1160 (1993)

37. T. Avitabile, F. Marano, F. Castiglione, C. Bucolo, M. Cro, L. Ambrosio, C. Ferrauto, A. Reibaldi, Biocompatibility and biodegradation of intravitreal hyaluronan implants in rabbits. Biomaterials **22**(3), 195–200 (2001)

38. B. Zavan, V. Vindigni, S. Lepidi, I. Iacopetti, G. Avruscio, G. Abatangelo, R. Cortivo, Neoarteries grown in vivo using a tissue-engineered hyaluronan-based scaffold. FASEB J. **22**(8), 2853–2861 (2008)

39. V. Vindigni, R. Cortivo, L. Iacobellis, G. Abatangelo, B. Zavan, Hyaluronan benzyl ester as a scaffold for tissue engineering. Int. J. Mol. Sci. **10**(7), 2972–2985 (2009)

40. G. Pasquinelli, C. Orrico, L. Foroni, F. Bonafè, M. Carboni, C. Guarnieri, S. Raimondo, C. Penna, S. Geuna, P. Pagliaro, A. Freyrie, A. Stella, C.M. Caldarera, C. Muscari, Mesenchymal stem cell interaction with a non-woven hyaluronan-based scaffold suitable for tissue repair. J. Anat. **213**(5), 520–530 (2008)
41. B. Grigolo, G. Lisignoli, G. Desando, C. Cavallo, E. Marconi, M. Tschon, G. Giavaresi, M. Fini, R. Giardino, A. Facchini, Osteoarthritis treated with mesenchymal stem cells on hyaluronan-based scaffold in rabbit. Tissue Eng. Part C: Methods **15**(4), 647–658 (2009)
42. P.S. Chan, J.P. Caron, G.J.M. Rosa, M.W. Orth, Glucosamine and chondroitin sulfate regulate gene expression and synthesis of nitric oxide and prostaglandin E2 in articular cartilage explants. Osteoarthritis Cartilage **13**(5), 387–394 (2005)
43. P. Du Souich, A.G. García, J. Vergés, E. Montell, Immunomodulatory and anti-inflammatory effects of chondroitin sulphate. J. Cell Mol. Med. **13**(8a), 1451–1463 (2009)
44. C. Malavaki, S. Mizumoto, N. Karamanos, K. Sugahara, Recent advances in the structural study of functional chondroitin sulfate and dermatan sulfate in health and disease. Connect. Tissue Res. **49**(3), 133–139 (2008)
45. K.R. Kirker, Y. Luo, J.H. Nielson, J. Shelby, G.D. Prestwich, Glycosaminoglycan hydrogel films as bio-interactive dressings for wound healing. Biomaterials **23**(17), 3661–3671 (2002)
46. K. Kojima, Y. Okamoto, K. Kojima, K. Miyatake, H. Fujise, Y. Shigemasa, S. Minami, Effects of chitin and chitosan on collagen synthesis in wound healing. J. Vet. Med. Sci. **66**(12), 1595–1598 (2004)
47. B.-M. Min, S.W. Lee, J.N. Lim, Y. You, T.S. Lee, P.H. Kang, W.H. Park, Chitin and chitosan nanofibers: electrospinning of chitin and deacetylation of chitin nanofibers. Polymer **45**(21), 7137–7142 (2004)
48. H.K. Noh, S.W. Lee, J.-M. Kim, J.-E. Oh, K.-H. Kim, C.-P. Chung, S.-C. Choi, W.H. Park, B.-M. Min, Electrospinning of chitin nanofibers: degradation behavior and cellular response to normal human keratinocytes and fibroblasts. Biomaterials **27**(21), 3934–3944 (2006)
49. C.K.S. Pillai, W. Paul, C.P. Sharma, Chitin and chitosan polymers: chemistry, solubility and fiber formation. Prog. Polym. Sci. **34**(7), 641–678 (2009)
50. T. Chandy, C.P. Sharma, Chitosan–as a biomaterial. Biomater. Artif. Cells Artif. Organs **18**(1), 1–24 (1990)
51. R. Muzzarelli, G. Biagini, A. Pugnaloni, O. Filippini, V. Baldassarre, C. Castaldini, C. Rizzoli, Reconstruction of parodontal tissue with chitosan. Biomaterials **10**(9), 598–603 (1989)
52. G. Biagini, A. Bertani, R. Muzzarelli, A. Damadei, G. DiBenedetto, A. Belligolli, G. Riccotti, C. Zucchini, C. Rizzoli, Wound management with N-carboxybutyl chitosan. Biomaterials **12**(3), 281–286 (1991)
53. M.P. Ribeiro, A. Espiga, D. Silva, P. Baptista, J. Henriques, C. Ferreira, J.C. Silva, J.P. Borges, E. Pires, P. Chaves, I.J. Correia, Development of a new chitosan hydrogel for wound dressing. Wound Repair Regeneration **17**(6), 817–824 (2009)
54. M. Ignatova, N. Manolova, N. Markova, I. Rashkov, Electrospun non-woven nanofibrous hybrid mats based on chitosan and PLA for wound-dressing applications. Macromol. Biosci. **9**(1), 102–111 (2009)
55. L. Wu, H. Li, S. Li, X. Li, X. Yuan, X. Li, Y. Zhang, Composite fibrous membranes of PLGA and chitosan prepared by coelectrospinning and coaxial electrospinning. J. Biomed. Mater. Res., Part A **92A**(2), 563–574 (2010)
56. B.-S. Liu, C.-H. Yao, S.-S. Fang, Evaluation of a non-woven fabric coated with a chitosan bi-layer composite for wound dressing. Macromol. Biosci. **8**(5), 432–440 (2008)
57. W. Wang, S. Itoh, K. Konno, T. Kikkawa, S. Ichinose, K. Sakai, T. Ohkuma, K. Watabe, Effects of Schwann cell alignment along the oriented electrospun chitosan nanofibers on nerve regeneration. J. Biomed. Mater. Res., Part A **91A**(4), 994–1005 (2009)
58. J.-Z. Wang, X.-B. Huang, J. Xiao, W.-T. Yu, W. Wang, W.-Y. Xie, Y. Zhang, X.-J. Ma, Hydro-spinning: a novel technology for making alginate/chitosan fibrous scaffold. J. Biomed. Mater. Res. A **93**(3), 910–919 (2010)

59. D.F. Williams, S.P. Zhong, Biodeterioration/biodegradation of polymeric medical devices in situ. Int. Biodeterior. Biodegradation **34**(2), 95–130 (1994)
60. J.W. Coleman, Nitric oxide in immunity and inflammation. Int. Immunopharmacol. **1**(8), 1397–1406 (2001)
61. R.L. Reis, J.S. Román, *Biodegradable Systems in Tissue Engineering and Regenerative Medicine* (CRC Press, Boca Raton, 2004)
62. R. Chandra, R. Rustgi, Biodegradable polymers. Prog. Polym. Sci. **23**(7), 1273–1335 (1998)
63. H. Tsuji, H. Muramatsu, Blends of aliphatic polyesters: V non-enzymatic and enzymatic hydrolysis of blends from hydrophobic poly(l-lactide) and hydrophilic poly(vinyl alcohol). Polym. Degrad. Stab. **71**(3), 403–413 (2001)
64. L. Liu, S. Li, H. Garreau, M. Vert, Selective enzymatic degradations of poly(l-lactide) and poly(ε-caprolactone) blend films. Biomacromolecules **1**(3), 350–359 (2000)
65. Z. Gan, Q. Liang, J. Zhang, X. Jing, Enzymatic degradation of poly(ε-caprolactone) film in phosphate buffer solution containing lipases. Polym. Degrad. Stab. **56**(2), 209–213 (1997)

Chapter 7
Current Applications of Biotextiles and Future Trends

Keywords Medical textiles · Implantable devices · Temporary support devices · Temporary barriers · Drug delivery devices · Scaffolds · Materiomics

7.1 Biomedical Applications of Fiber-Forming Resorbable Polymers

Some of the biomedical applications of each of the resorbable polymers have been discussed individually in the previous sections. Here, the general applications of fiber-forming bioresorbable polymers will be discussed. Table 7.1 shows the scope and range of biotextile devices currently used for medical applications.

The medical applications of biopolymers can be classified into three categories:

1. Permanently implanted devices
2. Temporary implants
3. Extracorporeal uses

The first category in particular requires materials which are biostable and last for at least the life expectancy of the patient. Temporary implants are required only as long as their function is needed inside the body. Once the injury has healed or the problem is resolved, they are no longer required. This is where the resorbable polymers prove to be useful since they are removed from the body without a second surgical intervention. Bioresorbable polymers have been explored for various applications such as resorbable sutures, barriers for surgical adhesions, scaffolds for tissue engineering, bone fixation devices, vascular grafts, stents, nerve growth conduits, ligament/tendon prostheses, anastomotic rings for intestinal surgery, implantable drug delivery systems, and many more. Based on their function, all these temporary applications for resorbable polymers can be classified into three main groups [1]:

C. R. Gajjar and M. W. King, *Resorbable Fiber-Forming Polymers for Biotextile Applications*, SpringerBriefs in Materials, DOI: 10.1007/978-3-319-08305-6_7, © The Author(s) 2014

Table 7.1 Current clinical applications of biotextile devices and their structure (adapted from [23])

Clinical application	Devices with biotextile component	Polymer type[a]	Fiber structure	Fabric structure
General surgery	Esophageal stent	Polyester (PET)	Monofilament	Braided
	Hernia repair mesh	Polyester (PET)	Monofilament	Warp knit
			Multifilament	Warp knit
		Polypropylene	Monofilament	Warp knit
			Multifilament	Warp knit
		PTFE	Expanded film	ePTFE membrane
		PGCL and polypropylene	Monofilaments	Warp knit
		Polylactide (PLA)	Monofilament	Self-fixation, pile component
	Patch	Polyester (PET)	Multifilament	Weft knit, double velour, nonwoven
		PTFE	Expanded film	ePTFE membrane
	Prolapse repair mesh	Polypropylene	Monofilament	Warp knit
		PTFE	Expanded film	ePTFE membrane
	Permanent suture	Nylon 6	Monofilament	–
			Multifilament	Braided
			Monofilament	–
		Polyester (PET)	Multifilament	Braided
		Polypropylene	Monofilament	–
		PBT-TMEG	Monofilament	–
		PVDF-HFP	Monofilament	–
		Silk	Multifilament	Braided
		Stainless steel	Monofilament	–
	Resorbable suture	Collagen (Catgut)	Monofilament	–
		Polydiaxonone (PDO)	Monofilament	–
		Polyglycolide (PGA)	Multifilament	Braided

(continued)

Table 7.1 (continued)

Clinical application	Devices with biotextile component	Polymer type[a]	Fiber structure	Fabric structure
		PGCL	Monofilament	–
		PLGA	Monofilament	–
		PGA-PCL-TMC-PLA	Multifilament	Braided
		PGA-PDO-TMC	Monofilament	–
		PGA-TMC	Monofilament	–
			Monofilament	–
Cardiovascular	Annuloplasty ring	Polyester (PET)	Multifilament	Weft knit
		PTFE	Multifilament	Weft knit
	Arteriovenous shunt	PTFE	Expanded film	ePTFE membrane
	Blood filter (ex vivo)	Polyester (PET)	Multifilament	Nonwoven
	Cardiac support device	Polyester (PET)	Multifilament	Knit
	Embolic vena cava filter	PTFE-FEP	Monofilament	Knotted
	Endovascular stent graft	Polyester (PET)	Multifilament	Woven
		PTFE	Expanded film	ePTFE membrane
	Heart valve sewing ring	Polyester (PET)	Multifilament	Weft knit
		PTFE	Multifilament	Weft knit
	Septal defect repair device	Polyester (PET)	Multifilament	Knit
		PTFE	Expanded film	ePTFE membrane
	Vascular prosthesis	Polyester (PET)	Multifilament	Warp knit, woven
		PTFE	Expanded film	ePTFE membrane
Dental	Dental reinforcing ribbon	UHMWPE	Multifilament	Leno weave
Neural	Nerve guide prosthesis	Polyglycolide (PGA)	Multifilament	Woven

(continued)

Table 7.1 (continued)

Clinical application	Devices with biotextile component	Polymer type[a]	Fiber structure	Fabric structure
Orthopedic	Bone graft/cement	Carbon fiber	Staple fiber/multifilament	Reinforced PMMA composite
	Ligament and tendon prosthesis	Polyester (PET)	Multifilament	Braided/woven/knitted
		PTFE	Expanded film	ePTFE membrane
		UHMWPE	Multifilament	Braided
	Spinal disk nucleus prosthesis	UHMWPE	Multifilament	Woven
	Spinal support	Carbon fiber	Staple fiber/multifilament	Reinforced PEEK composite
Skin and wound dressing	Skin graft	PLGA	Multifilament	Knitted
		Nylon	Multifilament	Knitted
		Nylon/collagen	Multifilament	Knitted velour
	Wound dressing	Cotton	Staple yarn	Woven
		UHMWPE/viscose Rayon and PET	Multifilament/staple fiber	Knitted/nonwoven composite
		Carboxymethylcellulose	Staple fiber	Nonwoven
		Cotton/viscose rayon	Staple yarn	Leno weave
		Viscose rayon/nylon	Multifilament	Knitted
		Polyester (PET)/viscose rayon	Staple fiber	Nonwoven composite
		Polypropylene/cellulose	Staple fiber/pulp fiber	Nonwoven composite
		Nylon	Multifilament	Leno weave
		Calcium or sodium alginate	Staple fiber	Nonwoven
Tissue regeneration	Tissue engineering scaffolds	PLA, PGA, PLGA, PLCL, PHB	Monofilament/multifilament	Nonwoven/woven/knitted—3D spacer/braided

[a] Abbreviations PLA poly(lactide), PET poly(ethylene terephthalate), PTFE polytetrafluoroethylene, FEP fluorinated ethylene propylene polymer, PVDF-HFP poly(vinylidene fluoride-co-hexafluoropropylene), PBT-TMEG poly(butylene terephthalate)-co-poly(tetramethylene ether glycol), PLGA poly(lactide-co-glycolide), PGCL poly(glycolide-co-caprolactone), PLCL poly(lactide-co-caprolactone), PCL polycaprolactone, TMC trimethylene carbonate, UHMWPE ultra-high molecular weight polyethylene, PEEK polyetheretherketone, PMMA polymethylmethacrylate, and PHB poly(hydroxy butyrate)

1. Temporary support devices
2. Temporary barriers
3. Drug delivery devices

Temporary support devices include sutures, bone fixation devices, tissue engineering scaffolds, vascular grafts, nerve guides, ligament prosthesis, etc. Resorbable polymers like PGA, poly(lactide) (PLA), poly(lactide-co-glycolide) (PLGA), and their copolymers were initially developed as suture materials. Much research has been undertaken to improve the strength of the fibers drawn from these resorbable polymers [2, 3]. Today, these polymers are commercially available as sutures, and they are FDA approved for this end use. Since PLA and PGA are stiff and inflexible polymers, initially they were used as multifilament braided structures, which provided them with good handling properties. However, for continuous suturing, braided structures with a non-smooth surface are not optimal. The rough surface causes elevated tissue drag, tissue trauma, and increases the risk of infection. Hence, more flexible polymers like PLGA and PDS were developed which could be used as monofilament sutures. Monofilament sutures have smoother surfaces and are easier to use for continuous suturing. There is minimal tissue drag and less tissue trauma, but on the other hand, there is lower knot security. Recently, polyhydroxyalkanoates (PHAs) and their copolymers have been studied for monofilament suture applications, and they have shown promising results. Over the past few years, many patents have been granted for biomedical applications of PHAs [4–11].

In recent years, resorbable polymers have found increasing use in orthopedic applications. Bone healing or regeneration is a dynamic process, beginning initially with no dimensional or mechanical stability and then gradually increasing in strength. Resorbable polymers with decreasing weight-bearing capacity over time are most suitable for bone healing. PLLA has been used widely for bone fixation devices because of its high strength and slow resorption rate [12]. For the same reason, resorbable polymers are an attractive choice for scaffolds for cell growth and tissue engineering applications. Scaffolds made from resorbable materials provide support to the growing cells and tissues, and finally disappear when they are no longer required. Almost all of the existing fiber-forming resorbable polymers have been explored for their potential as scaffold materials for tissue engineering applications [13, 14].

Temporary barriers are used to prevent tissue adhesion during the healing process. The barriers are not needed once healing is complete. Thus, resorbable polymers are suitable for this application. Both, bioresorbable films and fiber meshes have been explored as barrier devices [15, 16]. Drug delivery systems are also one of the most investigated areas of applications for resorbable polymers. However, fibrous or textile structures are not necessarily required for this application. Rather, surface-eroding polymers capable of forming hydrogels have been the target of most investigations for these drug delivery applications. Recently, there has been a growing trend to develop drug-eluting implants such as sutures and stents. Thus, the research for fiber-forming resorbable polymers capable of drug delivery has been the focus of several studies [17–22].

7.2 Future Trends

There are two factors driving the innovation and new applications for biotextiles. First, the clinical needs and practice, which are moving toward less invasive surgical techniques, such as endoscopy, and the use of resorbable biomaterials, which do not require explantation and hence require no second surgical operation; second, the increasing interest and commitment toward regenerative medicine, whereby resorbable scaffolds serve as temporary templates for the proliferation and regeneration of new viable tissues. Biotextiles are thinner, stronger, more flexible, porous, and lightweight structures compared to other biomaterials such as metals and ceramics, and they have already demonstrated their superiority to withstand the compression, tensile, shear, and bending forces that accompany folding, compaction, and delivery through a catheter.

The trend in the area of biomaterials is currently moving from permanent materials to, not just resorbable materials, but to functional bioresorbable materials. There is a growing research emphasis for developing combination polymers for biomedical applications. These are polymers in which the monomer contains multiple degradable groups. Unlike the copolymerization of different monomers, the molecular proximity of these groups yields functionally novel biomaterials. These materials have properties that cannot be obtained by single resorbable polymers or through simple copolymerization [24].

There is now a plethora of novel fiber-forming polymers in development, which have unique properties, such as shape memory, that are electroactive, elastomeric, rapidly resorbable, and responsive. When spun into fibers or as biotextiles, they can serve as biosensors, actuators, and drug delivery systems by responding to changes in temperature, pH, moisture level, and drug concentration. An example of one of the interesting developments is the 'smart' suture that can be used to seal difficult wounds where access is limited. Its self-knotting action occurs when it is heated a few degrees above normal body temperature. It uses a shape memory polymer that contains a hard segment and a switching segment, both having different thermal properties. The material forms a temporary shape at one temperature and a permanent shape at a higher temperature. On increasing the temperature, the suture material shrinks, creating a knot with just the right amount of tension on the surrounding tissue [25]. Bioactive sutures are being developed that would not only serve for wound closure but also show other functional properties like antimicrobial activity, improved cosmesis, and wound-healing properties. Novel bioactive suture materials have been developed by coating Vicryl® sutures with bioactive glass powder that enhances wound healing [26]. Drug-eluting coatings on the suture materials can impart superior functionality. For example, encapsulated antiseptic drugs can be released post-implantation to prevent infection at the wound site. Advanced surface modification techniques will lead to the development of biotextiles with immobilized biomolecules, growth factors, and antimicrobial agents. This can already be seen in the form of scaffolds with growth-inducing factors, also known as tissue engineering 'templates'.

With the growing numbers of different resorbable polymers with functional properties, there will be the need for non-conventional extrusion spinning and sterilization techniques to preserve the inherent characteristics and prevent premature degradation. New spinning techniques, like electrospinning and bicomponent spinning, have extended the range of manufactured fiber diameters from about 10 μm down to less than 50 nm and have also facilitated the spinning of fibers with a range of different cross-sectional shapes. These finer non-circular bicomponent fibers appear to provoke a different cellular response that as yet is not fully understood, but lends itself to fabricating implantable devices and tissue engineering scaffolds that have uniquely tunable cellular responses, rates of resorption, and drug delivery properties [27].

With the advancement in computational capabilities, the use of computational tools will facilitate the development and screening of new biomaterials. Mathematical modeling and computer simulations are being used to understand the structure–processing–property relationships for biopolymers. Such computational approaches would not only give better insights into the resorption mechanisms but would also enable the modulation of resorption properties by controlling the chemical composition and processing parameters. An integrated and holistic approach involving the disparate fields of material science, biology, and computational technology has been proposed for the development of new biomaterials. This system for the holistic study of biomaterials is now referred to as *materiomics* [28].

Conventional resorbable polymers like PLA and PGA are the preferred choice of biomaterials for wide variety of applications, primarily because they are FDA approved for various applications, and they have a safe and clinically proven track record. However, it should be realized that these are not the ultimate biomaterials that could be used for all applications. For example, PLA might not be the ideal candidate for tissue regeneration scaffolds, given the acidic by-products and autocatalytic degradation mechanism. Hence, there is a need to explore and commercialize other resorbable biomaterials. The search for stronger and more elastic biomaterials with wide range of resorption times has led to the development of novel bioresorbable materials in addition to conventional synthetic polymers. PHAs and their copolymers with collagen, chitin, or alginate components are just a few of the more promising materials under investigation.

In summary, the field of resorbable biotextiles will continue to grow as new synthetic polymers, existing and genetically engineered biopolymers, fiber spinning technologies, three-dimensional fabric constructions, coatings, and surface modification processes are introduced. Computational tools will facilitate the convergence of distinct fields of material science and biology leading to the development of new and improved biomaterials.

References

1. T. Hayashi, Biodegradable polymers for biomedical uses. Prog. Polym. Sci. **19**(4), 663–702 (1994)
2. J. Rose, High strength bioreabsorbable co-polymers, US 20080045627 A12008
3. J. Rose, High strength bioresorbables containing poly-glycolic acid, 745567425-Nov-2008 (2008)
4. I. Noda, Fibers, nonwoven fabrics and absorbent articles comprising a biodegradable polyhydroxyalkanoate comprising 3-hydroxyalkanoate and 3-hydroxyhexanoate, US614394711-Jan-2000 (2000)
5. S.F. Williams, Bioabsorbable, biocompatible polymers for tissue engineering, US651451504-Feb-2003 (2003)
6. S.F. Williams, D.P. Martin, T. Gerngross, D.M. Horowitz, Polyhydroxyalkanoates for in vivo applications, US7906135
7. S.F. Williams, D.P. Martin, F.A. Skraly, Medical devices and applications of polyhydroxyalkanoate polymers, US7553923
8. D.P. Martin, S. Rizk, A. Ahuja, S.F. Williams, Polyhydroxyalkanoate medical textiles and fibers, US8034270
9. G. Terenghi, P.-N. Mohanna, D.P. Martin, Polyhydroxyalkanoate nerve regeneration devices, US 2009/0209983 A1
10. S. Rizk, Non-curling polyhydroxyalkanoate sutures, US8084125
11. U. Gohs, B. Taendler, R. Vogel, D. Voigt, Nucleated poly-3-hydroxybutyric acid fiber for use in non-wovens, e.g. for medical implants, made from a mixture of high- and low-molecular wt. poly-acid and crystallised in the alpha and beta modification, DE102007000694-A105-Mar-2009 (2009)
12. B. Gupta, N. Revagade, J. Hilborn, Poly(lactic acid) fiber: an overview. Prog. Polym. Sci. **32**(4), 455–482 (2007)
13. M. Martina, D.W. Hutmacher, Biodegradable polymers applied in tissue engineering research: a review. Polym. Int. **56**(2), 145–157 (2007)
14. P.A. Gunatillake, R. Adhikari, Biodegradable synthetic polymers for tissue engineering, Eur. Cell Mater. **5**, 1–16; discussion 16, (2003)
15. H. Magnusson, T. Mathisen, Mesh implant with an interlocking knitted structure, 801684113-Sep-2011 (2011)
16. M. Therin, A. Meneghin, J.-L. Tayot, S. Montanari, Oxydized cellulose prosthesis, US 20070032805 A12007
17. M. Zilberman, R.C. Eberhart, Drug-eluting bioresorbable stents for various applications. Annu. Rev. Biomed. Eng. **8**(1), 153–180 (2006)
18. C. Di Mario, H. Griffiths, O. Goktekin, N. Peeters, J. Verbist, M. Bosiers, K. Deloose, B. Heublein, R. Rohde, V. Kasese, C. Ilsley, R. Erbel, Drug-eluting bioabsorbable magnesium stent. J. Interv. Cardiol. **17**(6), 391–395 (2004)
19. R.S. Schwartz, E.R. Edelman, A. Carter, N. Chronos, C. Rogers, K.A. Robinson, R. Waksman, J. Weinberger, R.L. Wilensky, D.N. Jensen, B.D. Zuckerman, R. Virmani, Drug-eluting stents in preclinical studies recommended evaluation from a consensus group. Circulation **106**(14), 1867–1873 (2002)
20. A.E. Deliaert, E. Van den Kerckhove, S. Tuinder, S. Fieuws, J.H. Sawor, M.A. Meesters-Caberg, R.R. van der Hulst, The effect of triclosan-coated sutures in wound healing. A double blind randomised prospective pilot study. J. Plast., Reconstr. Aesthetic Surg. **62**(6), 771–773 (2009)
21. B. Pasternak, M. Rehn, L. Andersen, M. Ågren, A.-M. Heegaard, P. Tengvall, P. Aspenberg, Doxycycline-coated sutures improve mechanical strength of intestinal anastomoses. Int. J. Colorectal Dis. **23**(3), 271–276 (2008)
22. R. Zurita, J. Puiggalí, A. Rodríguez-Galán, Triclosan release from coated polyglycolide threads. Macromol. Biosci. **6**(1), 58–69 (2005)

23. M.W. King, S. Chung, Medical fibers and biotextiles, in *Biomaterials Science: An Introduciton to Materials In Medicine*, 3rd edn., ed. by B.D. Ratner, A.S. Hoffman, F.J. Schoen, J.E. Lemons (Academic Press, Waltham, MA, 2012), pp. 301–320
24. B.D. Ulery, L.S. Nair, C.T. Laurencin, Biomedical applications of biodegradable polymers. J. Polym. Sci., Part B: Polym. Phys. **49**(12), 832–864 (2011)
25. A. Lendlein, R. Langer, Biodegradable, elastic shape-memory polymers for potential biomedical applications. Science **296**(5573), 1673–1676 (2002)
26. O. Bretcanu, E. Verné, L. Borello, A. Boccaccini, Bioactivity of degradable polymer sutures coated with bioactive glass. J. Mater. Sci. Mater. Med. **15**(8), 893–899 (2004)
27. S. Chung, M.P. Gamcsik, M.W. King, Novel scaffold design with multi-grooved PLA fibers. Biomed. Mater. **6**(4), 045001 (2011)
28. S.W. Cranford, J. de Boer, C. van Blitterswijk, M.J. Buehler, Materiomics: an-omics approach to biomaterials research. Adv. Mater. **25**(6), 802–824 (2013)

Index

C. R. Gajjar and M. W. King, *Resorbable Fiber-Forming Polymers*
for Biotextile Applications, SpringerBriefs in Materials,
DOI: 10.1007/978-3-319-08305-6, © The Author(s) 2014